ADVANCED AUDIO PRODUCTION TECHNIQUES

ADVANCED AUDIO PRODUCTION TECHNIQUES _____

Ty Ford

Focal Press
Boston   London

Focal Press is an imprint of Butterworth–Heinemann.

Library of Congress Cataloging-in-Publication Data
Ford, Ty.
  Advanced audio production techniques / Ty Ford.
    p. cm.
  Includes index.
  ISBN 0-240-80082-6
  1. Sound—Recording and reproducing.   I. Title.
TK7881.4.F67   1993   93–3353
621.389'3—dc20   CIP

British Library Cataloguing-in-Publication Data
A catalogue record for this book is available from the British Library.

Butterworth–Heinemann
313 Washington Street
Newton, MA 02158–1626

10 9 8 7 6 5 4 3

Printed in the United States of America

This book is dedicated to my parents Julia Malone Ford and Stephen Highland Ford. Without their passion for life and nurturing love this book would never have been possible.

To Helen Shortal, who rescued me with her caring and sense of humor when the task of getting the job done seemed too great.

And to author Michael Keith for his interest in my writing, and his encouragement to write this book.

Contents

Preface

I was first attracted to audio and electronics at the age of nine, and I have been fascinated by them ever since. I feel very fortunate that I have been able to make a living working in and writing about these fields. Much of the writing I have done has been aimed at demystifying the art, craft, and science of audio so that it can be put into the hands of more people.

Several years ago, when I began writing "Producer's File" for *Radio World*, it occurred to me that an information bottleneck had developed in the audio production industry. On one side were the designers and manufacturers who were making the equipment. On the other side were people who were using the equipment, an increasing number of whom did not have technical backgrounds. These new users needed to know how to connect and operate the equipment and how to get the most out of it.

Unfortunately, much of this information was buried in technical jargon or never addressed at all. Thus, nontechnical readers were forced to learn a new language before they could gain access to even part of the knowledge. In response to that need, I have made a serious effort with this book to use words, language, and diagrams that communicate the knowledge without demanding that the reader be a master of the technical language. That is not to say that this book does not contain technical language. It does. Most of the time, however, I have explained the concepts before applying the technical vocabulary. I also relied heavily on the use of visual language so that video and film professionals, who normally have a higher-than-average visual acuity, will find this book an easy source of information.

It is a book for people who produce audio and want to do it better, regardless of their level of expertise. It is not just about how the pieces of a system can be connected; it is also about what to do with them after you have connected them. It is about the process of combining electronics, acoustics, and art with communications concepts to produce powerful and compelling productions.

Do not delude yourself into believing that knowing everything in this book will make you a world-class production expert. The best you can hope for is to see the spectrum of possibilities more clearly. It would be a disservice to claim that this or any other text covers everything you need to know. The true expert never stops learning. In fact, the expert who stops learning ceases to be an expert. I hope that curiosity will be as good a friend to you as it has been to me.

ADVANCED AUDIO PRODUCTION TECHNIQUES

Developing Golden Ears

Most of us take sound for granted. We hear sound constantly, even when we are asleep. If placed in an anechoic chamber, a specially designed room that absorbs all sound, you would still hear the sounds your body makes as those vibrations make their way along your skeleton to the delicate structures in the middle and inner ear.

The information in this chapter and the ones that follow will provide you with the high level of understanding you need to become a professional audio producer. It will change the way you think about sound and the way you listen. You will learn each of the individual elements that, when combined, provide a solid base that will give you a decided advantage in the professional audio industry. Although some of the information may seem almost too simple and basic, it is important to realize that it is your command of the basics that makes the difference. To that end, each chapter provides insights and examples of the thought process used to determine how a production should sound.

Just as the number of colors on a painter's palette and the quality of his brushes and canvas are not a sign of a great painter, memorizing every detail of this book and having access to the best production studio in the world may not make you a great producer. The professional audio producer constantly processes theoretical, practical, and experiential information, judging the outcome with her ears. The first step in knowing how to combine a given collection of sounds successfully so that the final product sounds the way you intended is learning how to listen.

HEARING

Hearing and listening are different. Hearing, the physical ability to perceive sound, becomes listening when sound is filtered, manipulated, and processed by the brain. Because the best audio producers are highly aware of the sound with which they are working, they are better equipped to make judgments about its use.

Most of us hear sound that varies in frequency from 40 hertz (Hz) to nearly 15 kilohertz (kHz). The very low frequency sounds emitted by industrial construction equipment and the bass drum or kick drum in a drummer's drum kit are in the 20- to 100-Hz range. Most of us are familiar with 60-cycle hum, the sound that is emitted from poorly connected audio equipment. Many can hear the high-pitched whine a television set emits when it is on. Depending on the make and model, that frequency ranges from 15.5 to 15.75 kHz.

It is not unusual to find gifted audio producers and engineers who can hear frequencies up to 18 kHz or more. People with this kind of extended hearing range are often said to have *golden ears*. Having golden ears alone, however, does not guarantee that you will become a gifted producer. In addition to having exceptional hearing, you must also be able to understand what you hear.

The relatively flat line in Figure 1.1 indicates that all frequencies from 20 Hz to 20 kHz, which has come to be known as the *audible spectrum*, are present at exactly the same volume. Because all of the frequencies exist at the same volume, this graph shows a linear or "flat" response. The actual sound, called *pink noise*, is perceived as a rushing or hissing sound.

A circuit such as an amplifier, preamplifier, or mixer is said to be *flat* or *linear* if it does not change, or "color," any of the signal that passes through it. In truth, most circuits do impart some small characteristics to the sound

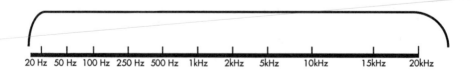

| 20 Hz | 50 Hz | 100 Hz | 250 Hz | 500 Hz | 1kHz | 2kHz | 5kHz | 10kHz | 15kHz | 20kHz |

Figure 1.1 A flat response curve. All frequencies from 20 Hz to 20 kHz exist at the same volume.

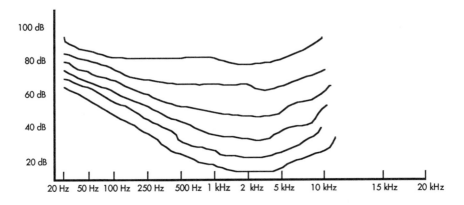

Figure 1.2 The Fletcher–Munson curve, showing the nonlinear nature of human hearing for the same sound perceived at six different volumes. Each of the six curves represents the equalization necessary to correct for the nonlinear nature of human hearing for a given playback volume.

that passes through them. To the finely trained producer's ear, knowledge of these subtle variations makes the difference between mundane and amazing recordings.

The Nonlinear Ear

To make things more challenging, unlike the flat response shown in Figure 1.1, human hearing is not linear. As shown in the Fletcher–Munson curve (Figure 1.2), the lower the volume of a particular audio source, the more we need to add low- and high-frequency content. This is why many consumer audio equipment makers put loudness controls on stereo equipment. At low listening levels, the loudness circuit compensates for our inability to hear lower and higher frequencies by increasing their volume relative to the middle frequencies. Many loudness circuits are designed so that their effect on the audio is automatically decreased as the volume is turned up. This keeps the bass and treble frequencies from becoming too loud at higher listening levels.

It is important to listen to your work over a variety of volume levels to ensure that all of the elements in the mix remain balanced. The more complex your production, the more important listening at different levels becomes. If your finished production is several minutes long, it is a good idea to listen the first time at a normal level. During your second listen, vary the volume above and below the normal level to make sure that even at moderate extremes your mix remains balanced.

LISTENING

Your trek toward production perfection will probably cause you to reexamine and change some of your listening habits. Most audiophiles go through a period in which they become fascinated with equalization (EQ). Armed with their first equalizer, they typically increase the bass or lower frequencies, then the treble or higher frequencies. Their favorite frequency response curves may be anything but flat.

The more you work as a producer, the more you come to appreciate linear response. As you become a better listener, you begin to notice that EQ changes more than just tonal balance. For example, analog equalizers achieve their effect by adding and canceling frequencies. As a by-product of this procedure, other changes in the audio occur that "smear" the sound, making it less distinct. In an effort to minimize this effect, the producer will choose a microphone with a frequency response curve likely to provide the appropriate sound with as little EQ as possible.

In a voice studio, it is not unusual to see different microphones used for male and female voices. The AKG 414 is often used on male voices, and the Neumann U87 condenser microphone is often used on female voices. Because the AKG 414 reproduces more high frequencies than the Neumann U87 and because the male voice has fewer high frequencies than the female voice, the 414 is often chosen to "brighten" the male voice, or increase its high frequency content. The Neumann U87 is used on the female voice to tone down the high frequencies so that both male and female voices have a more even response when heard together.

Professional Listening

Experienced audio producers listen differently than most people. In addition to content and performance continuity, they are sensitive to distortion, changing volume levels, masking effect, texture, and the ever-widening collection of processing effects found in most studios today.

Distortion

Distortion, in its purest definition, is anything that changes the original sound. Adding or subtracting bass or treble frequencies with an equalizer, for example, is distortion. Distortion also occurs when the audio level passing through a circuit or device exceeds the volume or signal level planned by the designer. When this occurs, large portions of the audio are chopped off, creating what some producers refer to as the *chain saw effect.*

Changing Volume Levels

Awareness of changing volume levels is important. If a musician, vocalist, or voice talent does not perform at a consistent level, the parts of the performance that dip in volume will not be heard. Playing, singing, or speaking too loudly for short durations is just as bad. In either case, the producer should ask the talent to perform with more consistent levels. If the problem continues, the producer can "ride gain" (manually control the level of the performance), use a compressor, limiter, or both, or do any combination of the three to ensure a consistent level. The operation of compressors and limiters is discussed fully in Chapter 5.

Masking Effect

Masking effect refers to the phenomenon of a sound or sounds that cover up, or mask, others. A positive example of masking is the use of music behind a narration track to cover up tape hiss or low-level background noises. Some producers record room tone, which is simply a recording of an open mike in a narration studio, to insert between and run underneath edited sections of narration. Room tone prevents the edits from being noticed.

Masking also can be undesirable. For example, a recorded guitar may sound perfect when recorded and listened to by itself. However, if other instruments are added to the recording, their combined acoustical energy may mask the sound of the guitar, preventing it from being heard clearly. In this situation, the producer has to make room for the guitar by lowering the volume of the masking instruments or by equalizing the other instruments, the guitar, or perhaps both so that their frequencies do not overlap each other to such a great degree. If the recording is in stereo, the producer can also solve the problem by placing the problem instruments so that they appear in different places from left to right across the stereo spectrum. The stereo spectrum is simply the soundstage that is created between the left and right speakers.

Texture

Texture is a question of degree and judgment. Although the producer may want to separate the guitar from the keyboards, he may want to combine several keyboard parts to create a texture. Again, EQ and stereo placement play an important role in the formation of the texture. EQ can be used to make the different keyboard parts sound less distinct and more like each other. Sounds with an abundance of high frequencies can be equalized so that some of those high frequencies are rolled off, or turned down. Sounds

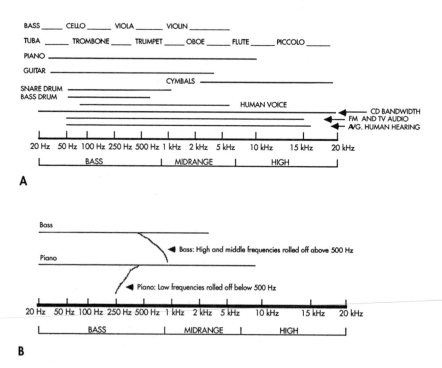

Figure 1.3 (A) Frequency ranges of common instruments before equalization. (B) Frequency ranges of bass and piano after equalization.

with an abundance of low frequencies may be equalized to roll off some of the low frequencies.

Figure 1.3A shows bass, drums, guitar, and piano and the parts of the frequency spectrum that these instruments occupy. Notice that many parts of the frequency spectrum are occupied by more than one instrument, particularly the 100-Hz to 5-kHz range. Without EQ, the mix of these instruments may contain an overabundance of energy between those frequencies that may make the mix sound muddy, or bassy. To prevent this buildup, the producer may decide to equalize or roll off the less important parts of each instrument's sound. As Figure 1.3B shows, the producer can roll off the lower frequencies of a piano part and combine it with a bass part that has had all of the midrange and high parts rolled off. The amount of EQ used determines the amount of separation among the instruments. After EQ, it is not unusual to find that each element of the mix sounds very unnatural by itself but great in the mix.

Processing Effects

The control of space made possible by the great number of delay and reverb processors is also part of the producer's domain. The best of these devices sound very natural, giving the listener the impression that the sound is occurring in a very small space, in a huge cathedral, or in a canyon. These devices can also be adjusted to produce very bizarre-sounding echoes and reverberations. It is the producer's job to be in touch with both the natural and the bizarre. As later chapters explain, the best way to learn is to experiment.

A good producer develops the ability to focus in on individual sounds within a mix. Similarly, some music appreciation courses you may have encountered require that you concentrate on various instruments within a piece of music. This kind of concentration resembles the concentration that the conductor of an orchestra must have. Like the conductor, you must develop your ability to shift attention from the smallest detail to the overall sound of the mix in a split second.

Unlike the conductor, however, your listening must include the ability to recognize the distortion caused by malfunctioning or misadjusted equipment and its effect on any audio that passes through it. You also need to know how to isolate the source or sources of signal problems and must be able to correct them.

As your ability to listen improves, you will be able to identify EQ, delay, reverb, echo, phasing, gain reduction, and the entire menu of acoustic processing and will develop an acute awareness of the frequency spectrum of audible sound. Learning to identify the frequencies of a sound by simply hearing it is very important. Following that, you will learn how and when to "sculpt" existing sounds with EQ so that each element of a mix can be used more effectively. Later, you will learn how to fine-tune this ability. In the beginning, this ear training may seem very difficult. First, become familiar with each type of effects circuit separately. If you are now working in a studio, use downtime or time between sessions to increase your familiarity with the effects on hand. Begin by learning how every knob and switch on every piece of equipment changes the sound of the audio passing through the circuit.

You may find portions of the manual for the device helpful, but do not count on it. Although some manuals are very well written, others offer little practical help. As you gain listening experience, you also gain a better understanding of what you are hearing and how to use that understanding to achieve the desired effect. For now, remember that the more you practice, the better you get. Over time, this accumulated awareness becomes your menu of possibilities—your bag of tricks.

USING POINTS OF REFERENCE

Experienced music producers, even some whose hearing is quite good, establish aural points of reference by comparing their mixes to other recordings of a similar nature (for example, similar types of music or similar voice-overs). By doing this, they develop a point of reference to prove to themselves that their work is balanced.

In a music project, the procedure is as simple as comparing a finished recording with well-known records or compact discs (CDs). So great are the differences between records, cassettes, and CDs that producers often make different master mixes for each medium. Incidentally, the more you experiment with comparative listening, the more you'll realize how different and subjective a producer's decisions are.

Voice-over music production for commercials or industrial narrations can also benefit from comparative listening. If you are working in a broadcast facility and a radio spot or television audio track you hear on the air seems to sound louder, or "punches through" better, than others around it, find out why. Listen to the master tape in the production studio. It may have been produced with an exaggerated amount of middle or high frequencies. If so, it will sound brighter than normal. Listen also for the amount of low frequencies relative to your own work.

Many producers who work in radio and television start by rolling off a certain amount of both highs and lows. This kind of equalizing reduces the bandwidth of the audio so that it takes up less space on the frequency spectrum. Because the sound takes up less space on the frequency spectrum, compressors and limiters don't have to work as hard as they do with full-spectrum audio.

Generally speaking, compressors automatically increase the lower levels of a signal, and limiters keep levels from becoming too loud. Most radio and television stations compress *and* limit their audio before it's broadcast. They do this to make the audio louder and to keep it from exceeding the legal limits of loudness established by the Federal Communications Commission (FCC). A good way to see the effects of compression and limiting is to feed the broadcast audio into a device with meters, such as a mixer or a tape machine. Watch the level meters for the total amount of fluctuation from the loudest to the quietest portions of the audio. If the needle or meter just seems to hang in one place with very little deviation, you know that the audio has been compressed, limited, or both.

Most producers, aware that broadcast processing involves EQ, compression, and limiting, try to process their audio so that it is less affected by broadcast processing. They do this to preserve the original qualities of the sound they have worked so hard to produce. Over time, producers find EQ, compressor, and limiter adjustments that become a starting point for their finished mixes and part of their own characteristic sound.

PRACTICING VISUALIZATION

Practicing visualization is an important step in developing the kind of listening skills you will need to be competitive as a producer. You can start by analyzing your own work or that of others. Some people find they can concentrate better on what they are hearing if they close their eyes.

Sit or stand in front of and between the speakers so that you form a triangle with them. The distance between the speakers should be slightly less than the distance between you and either speaker. As you listen to stereo music, for example, try to identify from what parts of the stereo spectrum different sounds seem to come. Concentrate on one sound and see if it stays in the same place or moves. Listen for sounds that only come out of one speaker. Do all of the sounds seem to be occurring in the same room, or do some sounds seem to be very far away in a big room while others seem very close and sound like they are being made in a small room? Is the space in which the music is occurring a real space, or does it sound manufactured by reverb, echo, and delay?

Combine the stereo signal to mono, if possible, and notice what effect that has on the size of the space. Do the instruments and vocals maintain their relative volume levels, or do they change? The more you listen, the more you learn.

You may find it helpful to use stereo headphones, which provide an isolated listening environment as compared to a set of speakers. Be aware that it's not a good idea to rely solely on headphones when mixing audio, especially if the finished production will be heard over speakers. The extreme separation and lack of fidelity of most headphones usually result in a mix that sounds great in the headphones but terrible when heard on studio monitors or other speakers.

If you're working in a properly designed production studio, the effects of placement on the stereo spectrum should be easy to hear. You may notice that some sounds are louder than others, and that different sounds seem to be in spaces of different sizes. In contemporary recordings, the drums are often mixed so that they appear to be played in the middle of a very large empty room and at some distance from the microphones. In reality, most of the microphones are often only several inches away from the drums. Solo instruments and vocals are usually mixed to sound much closer to the speakers, more "up front."

It is only after you get a firm understanding of the true dimensions of the landscape that you can begin to use it effectively to stage your own productions. Keep in mind, too, that as new audio effects devices are invented, the possibilities also increase.

Because most of today's audio is stereo, your ability to use the stereo spectrum is extremely important. Begin by thinking about sound as occurring in four dimensions: height, or volume; width, or stereo separation; depth, or distance from a sound source; and time. Time refers to when a sound is heard relative to others and when reflections of a sound are heard relative to the original sound. Accepting sound as something that has physical dimensions and occurs in time as well as space allows sound to be visualized. Once you begin to see sound, you can better judge where and how to place the elements of your production.

STARTING WITH A CONCEPT

Each production you work on should start with a concept. If the spoken word is part of the production, the copy must be as strong as the concept. The next steps involve exploring the possibilities created by various combinations of elements. For example, the stereo spectrum, time domain effects, EQ, and timing between elements can all be used to separate or combine sounds in a mix. Decisions as to which path to take should be guided by a well-thought-out original concept. Without such a concept, the number of variables quickly becomes mind-boggling, and the project takes a lot longer than it should. This not to say that some experimentation should not be allowed, but an experienced producer knows where to draw the line between accepting unexpected ideas that strengthen the concept and wasting time.

HEARING LOSS

Hearing loss can be categorized as short-term or long-term loss. Short-term hearing loss occurs when you are briefly exposed to loud sounds, such as those in a machine shop, from a gas-powered lawn mower, or from a concert or night club sound system. With this type of exposure, your hearing threshold temporarily shifts upward, making quieter sounds hard to hear or inaudible. The threshold normally returns to its previous level overnight or within a day or two. Although research shows that most people's hearing is relatively resilient to these sorts of brief exposures, there are also documented cases in which one single exposure has caused a permanent upward threshold shift, or permanent hearing loss.

Most producers protect their hearing by not exposing themselves to overly loud sound for extended lengths of time. Even the short-term hearing loss that occurs during a recording session of more than 6 hours can negatively affect a producer's hearing. After exposure to even moderately high volume audio for that period of time, most people suffer some temporary high-frequency hearing loss. As a result, the tendency is to overcompensate by adding more high frequencies than are really needed.

If you suspect that your threshold has shifted upward, it is a good idea to put off making any important mixing decisions until your threshold returns to normal. Otherwise you can overaccentuate the high frequencies, resulting in a production that is much too bright.

As we get older, most of us start losing the top and bottom of the audible spectrum. If you haven't taken a hearing test in more than 5 years, consider taking one at your local medical center or through a specialist to discover the exact range of your hearing.

You can make an informal test of your high-frequency hearing ability by turning on a standard television set. If you can't hear the high-frequency whine of the horizontal oscillator—a little over 15 kHz—face the fact that there are parts of the frequency spectrum that you just can't hear.

There is another practical way to check your hearing comparatively. Listen to music with a lot of high-frequency content (for example, cymbals or synthesizers) through a good system with an equalizer. Use the calibrated equalizer to vary the frequency response of the music. Unless there are major deficiencies in the system or your hearing, you should be able to hear EQ changes of more than 2 or 3 decibels (dB) anywhere between 40 Hz and 16 kHz. The decibel is the standard unit for the measurement of sound levels.

The best monitor systems are able to reproduce a range of at least 40 Hz to nearly 20 kHz. If you can hear those extremes without increasing the gain of the monitor amp, you have great hearing.

Even if your high-frequency hearing is not what it should be, don't give up. If a hearing test shows that your hearing is off at very high or very low frequencies, it is not the end of the world. In some cases, dysfunctional hearing can be repaired by surgery. If, however, your hearing loss is due to overexposure to very loud sounds, your hearing may not be restored.

To most producers, the ability to hear is their life. They are careful to limit their exposure to loud sounds to very short periods of time. Music mixing session monitor levels can often become harmful when they are too loud, even over a short period of time. The smart music producer uses earplugs or simply leaves the room when the client wants to hear a mix at high volume.

An early sign that you are overexposing yourself to overly loud audio is a ringing in the ears. The condition, called *tinnitus,* affects at least 30 million people in the United States. Although much research is being done to discover a cure, for many people there is no way to reverse the condition. If, in the quiet of a well-insulated recording studio, you still hear a constant ringing, buzzing, or rushing sound in either or both ears, it is probably tinnitus. You may want to check with an ear, nose, and throat doctor or an audiologist to determine the specifics of your condition. A call to the hearing and speech department of the local hospital or university may provide you with information about tinnitus support groups that meet regularly to discuss ways of coping with the problem.

DEVELOPING CHOPS

In addition to all of the increased mental processing required to develop your listening, to be a successful producer you also need to be able to push a lot of buttons, turn a lot of knobs, and move a lot of faders (linear volume controls) at precisely the correct rate and at just the right time. This takes practice and is called *developing chops.*

Don't be discouraged if you don't get it right the first time. Great mixes take time and usually remain great. Mixes that aren't great are like unrefrigerated fish; they're not so bad the first day, but they rapidly get worse. If your gut tells you something is wrong with the mix, your gut is probably right. Fix it before it gets out of your hands, or it will come back to haunt you.

Advanced Listening

2

In Chapter 1, the emphasis was on understanding how hearing works and focusing or mentally processing your hearing so that you can become an effective listener. This chapter adds to that knowledge by examining the listening environment more closely. You will learn how the acoustics of the control room, studio, and monitoring system affect your sound. You will also learn simple procedures that will allow you to determine quickly the source of many problems and to correct them.

Be aware that your production may be compromised by several problems occurring simultaneously. Finding the cause of one obvious problem and correcting it may uncover other, less-noticeable problems, or it may make those problems more noticeable. As you continue to learn about the equipment and its environment, interactions that first seemed mysterious will become more obvious. It is also important to know that the rules change as technology changes. Accepted practices of 5 years ago may not apply to the work you do today. It is your responsibility to know the rules so that you may question them when they don't appear to work.

STUDIO AND CONTROL ROOM ACOUSTICS

When explaining the acoustics of an environment, it is sometimes helpful to visualize sound pouring out of speakers as though it were water or a visible gas. If we were able to see sound, we would see eddies, dead spots, obvious reflections, peaks, and nulls. Because we cannot see these turbulences with the eye, we instead make measurements with calibrated microphones and a variety of noise and tone sources. If you have not listened intently before, try walking around the room while playing a familiar voice track, voice-over, music track, or other recording. As you move around the room, notice how the volumes of some frequencies vary. The location of the monitors, the shape of the room, angle of the walls and ceiling, and placement of reflective and absorptive materials all play a part in how much the sound changes between the time it leaves the monitors and when it reaches your ears.

Surfaces that are hard and smooth, like glass, reflect high frequencies much better than soft or porous surfaces. Early theorists in room acoustics sought to reduce reflected sound by installing carpeting or drapes on the walls of the studio. Unfortunately, these materials reduced a disproportionate amount of middle and high frequencies while having little effect on the low frequencies. As a result, accurate monitoring was difficult due to the bass-heavy imbalance of reflected sound. Attempts to solve the problem resulted in the building of bass traps. Slotted walls were designed with chambers behind them that absorbed the necessary amount of low frequencies.

Further developments in the study of acoustics have resulted in the live end–dead end (LEDE) theory. Simply put, the rear of the room, behind the producer, is the dead end. It is designed to absorb as much sound over as wide a range of frequencies as possible. This treatment prevents the sound leaving the monitors from reflecting off the back wall of the studio and into the producer's ears. The front of the room is the live end because that is where the monitors are.

Because most environments (living rooms, cars, auditoriums) reflect at least some sound, most designers shy away from totally dead control rooms or studios. Reverb and echo effects, which sound just right when produced in a dead room, can sound overdone when played back in a more live, or more reflective, environment.

Another aspect of acoustic environments that must be taken into consideration is their varying frequency response at different listening levels. Due to the structure, size, and shape of the control room, a monitor system's frequency response may be somewhat flat at one playback level, but resonant at another. The sounds you worked so hard to create in the production studio may sound terrible when played back at louder or softer levels.

MONITORS

Many factors must be considered when setting up a monitoring system. Theoretically, all studios should be set up so that the same tape, CD, or other source sounds the same when played back in any studio. Due to the many different makes and models of equipment; the different size, space, and construction methods for the studios and control rooms; and the placement of monitors in studios and control rooms, it is exceedingly difficult to achieve an absolute standard. There are just too many variables.

Unless you have and are prepared to use the proper equipment to measure the room and make accurate adjustments, it is better to consider contracting with someone with studio design experience. If lack of budget precludes the hiring of a designer, the following basic considerations will prevent you from making serious mistakes.

Monitor Coloration

Even the best monitoring systems impart their own characteristics to the audio they reproduce. The term *reference monitors* is often used by manufacturers to position their speakers as industry standards. A quick check of most professional reference monitors quickly shows that there is quite a variation in their frequency response. Typically, most monitors with larger bass speakers reproduce lower frequencies more effectively than monitors with smaller bass speakers. Other monitors may favor midrange or high frequencies. Because of this, it is not unusual for a production studio to have at least two sets of monitors and sometimes three.

The need for different monitoring systems arises in part from the lack of standardization and in part because mixing should be done on the speakers that most closely resemble the speakers over which the production will normally be heard. For example, although it's great to be able to hear your music mix on the larger monitors, where the full power of the bass shakes the floor, if you're recording a music group, your critical mixing should be done on smaller monitors that more closely resemble the frequency response of a listener's home or car speaker. If you know that the finished product will be played back over a cheap cassette machine attached to a slide projector or over the average 2-inch television speaker, you should check your mix on a similar speaker to make sure that all the low and midrange frequencies can still be heard.

If your final mix will be part of a multimedia display in a museum, it's a good idea to make a test mix and play it back at the site and on the actual equipment that will be used. You may find that the limitations of the play-back system and the acoustics of the space will force you to produce a mix that sounds wrong on your monitors but sounds great on the system over which it will actually be heard.

Monitor Installation

The same monitor may sound different depending on how and where it is mounted. A good rule to remember is that anytime you place a monitor on or near a wall, ceiling, floor, or other structural part of the studio without complete isolation, the interaction, or acoustical joining, that occurs between the monitor and the room will change the frequency response of the monitor system. In addition, the bigger the control room and the farther away you are from the monitors, the more the room plays a part in the sound.

Typically, the main monitors of a studio are either built into a wall, mounted on a wall, or suspended from the ceiling. When monitors are built into or mounted on a wall, it is important that they are physically and acoustically isolated from the wall. Without isolation, the wall or structure may resonate at certain frequencies. This resonance may add to or subtract from various frequencies of the sound produced by the monitors. That means that the mix you created in that particular studio may sound very different when played back over another system.

Even changing the monitors from being mounted in or on a flat wall to being mounted in the corners of the control room can noticeably affect the frequency response of the monitor system. You can expect to hear more bass frequencies when monitors are placed in the corner of a control room because the walls of the room act as extensions of the bass speakers. The monitors themselves do not produce more bass, but the walls that extend from the corner in which the monitors are mounted act like a megaphone, providing a large, continuous surface along which the bass frequencies can easily travel.

Corner-mounted monitors are affected even more if they are placed high in the corner against the ceiling or low in the corner near the floor. In both cases, the surfaces of the ceiling or floor become part of the megaphone, thereby altering the original sound of the monitor.

Monitors that are suspended from the ceiling by cables are not affected by the kind of acoustical joining found with built-in or wall-mounted monitors. However, they can transmit vibrations up the cables to the beams in the ceiling to which they are attached.

Because even well-isolated built-in and wall-mounted monitors mounted more than 6 feet away from the listening area at the console are susceptible to the acoustics of the room, many studios are equipped with a second set of monitors. They are usually smaller than the main monitors and are placed on top of the console, about 3 or 4 feet from the producer. Because these near-field monitors are closer to the producer, they are also less susceptible to the acoustics of the control room or studio.

Near-field monitors provide the producer with a valuable second perspective on a mix. They are usually connected to the same monitor amp so that the producer can quickly switch back and forth between monitors, comparing one to the other. This allows the producer to confirm that monitor coloration is not hiding a problem in the mix.

Moving larger monitors away from walls, ceilings, and floors so that they are closer to the listening area at the console may eliminate the need for additional near-field monitors, but this can create another problem. Most monitors are comprised of two or three speakers mounted in one enclosure. Each speaker reproduces a different range of frequencies and may be mounted up to several feet away from the other speakers on the face of the monitor. If the monitor is placed within 4 or 5 feet of the sweet spot, the producer hears the different frequencies coming from different locations on the face of the monitor, which can be very distracting.

Monitor Placement

No matter which kind of monitors you plan to use, a good starting point is to position the monitors so that they form an equilateral triangle with the area at the console at which the producer and engineer sit. The monitors should be turned slightly inward and aimed at that area of the console, which is called the *sweet spot* (Figure 2.1). A monaural source, when played

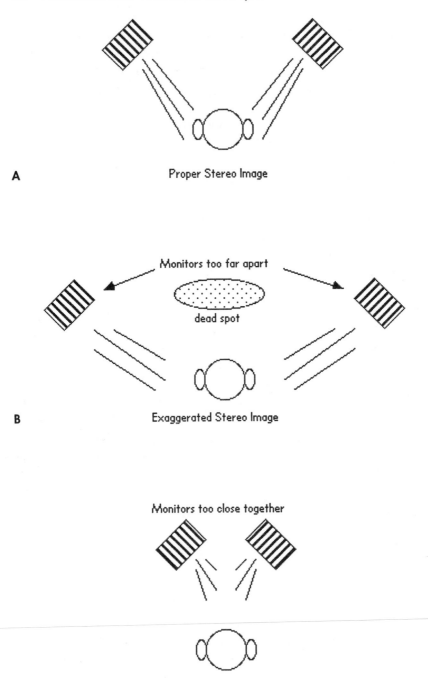

A   Proper Stereo Image

B   Exaggerated Stereo Image

C   Reduced Stereo Image

Figure 2.1   (A) Overhead view of monitors and sweet spot, showing equilateral triangle. (B) Monitors placed too far apart. (C) Monitors placed too close together.

through the monitors, should sound as though it originates from a point exactly between the monitors. The goal is to create the largest possible sweet spot.

The sweet spot is the area in which the sounds from a set of monitors intersect equally. Each speaker in a monitor has a characteristic called a dispersion angle. This angle, measured in degrees, is a measurement of the width of coverage of a speaker. Most high-frequency speakers have a narrow dispersion angle compared to midrange or low-frequency speakers. Therefore, as you move away from the sweet spot, the high frequencies are the first to fall off.

It may help you to consider the concept in terms of light. High frequencies project from a monitor the way a spotlight projects a beam of light. They are very directional. Low frequencies project more like a flood light and are much more omnidirectional.

Stereo Imaging

Monitor placement for stereo is critical. If the monitors are spaced too widely apart, such that the distance between the monitors is much greater than the distance between the sweet spot and either monitor, the producer may pan the elements of the mix too much to the center, which results in a mix that lacks stereo separation. If the monitors are too close to each other, such that the distance between the monitors is much less than the distance between the sweet spot and either monitor, the producer may tend to exaggerate the stereo panning to achieve stereo separation. Although these distortions in the stereo image may go undetected in the studio in which they were mixed, they will become apparent when the mix is heard on a correctly positioned monitoring system.

Console Reflectance

Console reflectance is an additional acoustical problem that occurs in studios with large consoles and in studios in which the monitors are positioned above the producer. In addition to hearing the sound that comes directly from the monitor, the producer also hears the delayed sound that is first reflected off the surface of the console (Figure 2.2 and 2.3). The irregular surface of the console can also change the tonal balance of the sound it reflects.

The reflectance problems with large consoles, which offer large irregular reflective surfaces, can be minimized by moving the monitors closer to

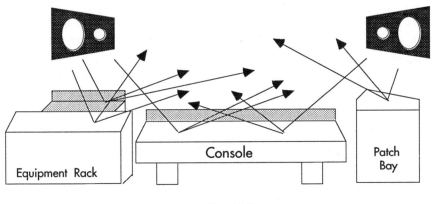

Front View

Figure 2.2 Typical reflectance of sound in a control room. The arrows and lines depict the audio paths.

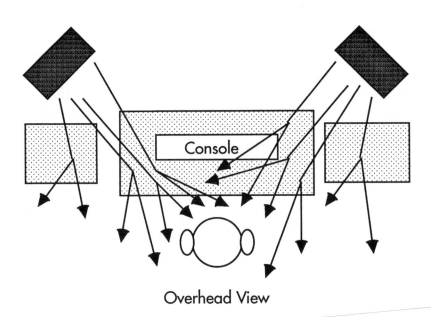

Overhead View

Figure 2.3 Console reflectance. The sound heard by the person at the console follows the pattern depicted by the arrows and lines.

the sweet spot. Repositioning monitors aimed down at the sweet spot from above so that they are at the same height as the producer's ears also reduces console reflectance.

RADIO AND TELEVISION REFERENCING

Audio producers in radio and television typically reference what they hear in the production studio to what that audio actually sounds like when broadcast. By doing so, they learn how line amps, signal processing, modulation, cart machines, and volume adjustments made by the board operator alter the EQ, presence, and sound of the master.

If you're doing production that runs on a number of stations, it is important to listen to several stations over several receivers. If your production sounds bad on one station, it may not necessarily mean that you have a problem. If it sounds bad on all the stations, you probably do.

In addition to the coloration of the monitor system—in this case, the power amp and speakers—you also have to account for what effect the receiver has on the signal. Cheap receivers, for example, may not faithfully reproduce what is being broadcast. They may introduce distortion of their own to the station's signal. In addition, if they are poorly designed, they may also prevent you from hearing the problems you would normally hear on a better receiver.

POLARITY AND PHASE CANCELLATION

The terms *polarity* and *phase* are often used interchangeably. Polarity is the simplest expression of phase. For example, a balanced or unbalanced circuit passes audio through two conductors: a positive, or high, conductor and a negative, or low, conductor. If these connections are reversed, their polarity, or phase, is changed by 180 degrees. Phase changes, however, may be anywhere between 0 and 360 degrees.

Stereo mixes should always be checked in mono for phase cancellation, especially if there is any chance that the material will be heard in mono. In any recording session at which more than one microphone is used, such as in a music recording session, a mono check should always be made. A mono check simply combines both stereo channels. Most consoles have a switch that allows monitoring to be changed from mono to stereo. This check ensures that the material you have mixed is in phase. Out-of-phase conditions most commonly occur when the polarity of one channel of a stereo signal is inadvertently reversed.

Experiment with phase cancellation by placing two similar mikes the same distance from one source and mixing them together in mono. Use the preamp or mixer to raise the level of the first mike to a normal operating level. Listen to what happens to the overall level as you raise the level of the other mike. If the mikes are in phase, the overall level will increase. If they are out of phase, the overall level will gradually decrease as the level of the second mike reaches that of the first.

If the two signals are identical but exactly opposite in phase, they will cancel completely, causing their combined volume to decrease. If one signal is greater than the other or if different mikes with different frequency responses are used, only partial phase cancellation is possible. Placing similar mikes at different distances from the source will also cause partial phase cancellation due to the difference in time required for the sound to reach each mike. In this case, complete cancellation probably will not be possible because, even though the signals from each mike may be very similar, they do not occur at exactly the same time. Phase relationships, then, are dependent on time, level, distance, and frequency. Understanding how these four variables interact and knowing which to alter in the pursuit of a certain sound are a function of experimentation and experience. Once you know what the variables are, cause-and-effect manipulation of the equipment becomes cumulative and makes more sense.

In a few situations, you may find that purposely reversing the phase of one or more channels of audio may improve your sound. Experimenting with this concept is easy if the console you are using has separate polarity switches for each of its inputs. The Greek letter *phi* (Ø) indicates that the switch or push button changes the polarity of the signal. If you have several mikes open at the same time while recording a group of people placed close together in the studio, change the polarity of each of the mike channels while listening for any improvement in the clarity of the audio. If you find that reversing the polarity of one or more of the mikes improves the sound in stereo, be sure to listen in mono if there is any possibility that the audio will be heard in mono.

This polarity reversal technique works because it uses phase cancellation to reduce the problem that occurs when one person's voice is picked up by several different mikes, including the one into which the person is speaking. Because the other mikes are farther away, the sound of the voice takes a few thousandths of a second longer to get to them. Instead of hearing the voice as directly picked up by one mike, the acoustic delay makes the voice sound as if the person is too far away from the mike. After a reasonable time experimenting, you may find that improving one person's sound makes another's worse, forcing you to compromise. Other remedies can be used to eliminate this problem. First, try placing the mikes farther apart. Second, use directional mikes rather than omnidirectional mikes, and arrange them so that each points at only one person and away from the others. As mentioned before, check for mono compatibility if there is any chance that the recorded audio will be played back in mono.

During a multitrack music session, this check also ensures that too much sound from a single source is not being picked up at slightly different times by different microphones due to their varying distance from the source of the sound. While most multiple-microphone recordings are affected somewhat by this phase canceling phenomenon, it is important to minimize it whenever possible.

Phase canceling can also occur in direct line inputs or outputs. Improperly installed studio equipment is occasionally the cause of the problem. If the high and low, or plus and minus, wires of one channel have been mistakenly reversed, the audio on that channel will be 180 degrees out of phase, and phase cancellation will occur. If two channels carrying the same audio are combined to mono and one is 180 degrees out of phase with the other, the signals will cancel, and no audio will be heard.

Being out of phase is not bad all the time. Phase shifts are responsible for many of the spatial sounds that stereo synthesizers, reverb devices, and chorus effects devices create. These circuits purposely change the phase any number of degrees to make the sound appear to move. As long as you're monitoring in stereo, there is no problem. When you switch to mono, however, the cancellation that occurs when the out-of-phase channel is combined with the in-phase channel can destroy the mix.

There is normally a reduction in volume in the monitors when out-of-phase stereo signals are combined to mono. In addition, the audio often sounds somewhat hollow. In the end, your mix should sound as good on a cheap system as it does on a good system, though it will obviously sound different.

LISTENING FOR NOISE

Many types of noise can compromise an otherwise good piece of production. Most types fall into three categories: acoustic, internal electronic, and external electronic. With practice you should be able to identify each type by ear and be able to isolate and eliminate it.

Acoustic Noise

Acoustic noise is any unwanted physical vibration that is picked up by a microphone or other type of transducer. Be aware that amp circuits, mike cables, and instrument cables can become microphonic. Although mostly a side effect of vacuum-tube circuits, any electronic device, because of either poor design or malfunction, may become microphonic. When this happens, the device conducts any physical vibrations and turns them into thumps, clicks, or other noises. If microphonic circuits are placed too near audio monitors, they can pick up the sound from the monitors and compromise the finished sound. Check for microphonic circuits by establishing normal operating levels all the way to and including the monitors and by tapping on all parts of the system with your finger. If you hear noise that corresponds to your tapping, have the circuit checked for any malfunction. If you cannot reduce the microphonic effect, make sure the circuit is well out of the way of the monitors and cannot be accidentally bumped.

Mike or instrument cables can generate noise by being slapped, tapped, or simply moved during the performance. Check for this by establishing normal operating levels and listening for noise as you move or slap the cables. If noise is apparent, either keep the cables still or use a higher-quality cable.

Voice Noises

When recording voice, popping *P*s are only one of many unwanted noises. Working a mike too close without a good pop filter can result in smaller eddies of breath being recorded as well. Be aware that many headphones lack the low-frequency response to detect popping. If you are monitoring a voice recording with headphones, make sure you check the playback over the studio monitors.

Mouth clicks should also be avoided. Mouth clicks are normally the result of a dry mouth or of too much saliva in the mouth. Most mouth clicks can be eliminated if the talent occasionally takes a drink of room-temperature water. Hot or cold liquids such as coffee, tea, or carbonated soft drinks should be avoided. Milk should also be avoided because it increases saliva flow in the mouth.

Internal Electronic Noise

Internal electronic noise is usually caused by bad or improper connections, malfunctioning circuitry, overdriven circuits, underdriven circuits, and bad ground connections. Real noise figures are often determined by using test equipment that can measure noise at levels well below the level shown on the console's output meters. Although listening for obvious noise is not as precise, it is a simple procedure that can reveal a lot about the system.

When the system is set up to record or play back, establish reasonable operating and monitor levels by feeding audio from your usual sources through the console. When the levels have been set, mark the volume or pot positions. Put all sources in their respective stop modes so they are not feeding audio to the console, but do not turn them off.

Leave the monitor volume at a good listening level, and reduce all the other pots or faders on the console to zero. Raise the console's master gain level to its marked operating level, and listen for hum or hiss. If you hear no noise and see no movement on the console's output meters, you can be fairly sure that there are no major problems in the master output section of the console.

Don't trust just your ears on this test; also check the meters. Noise that registers on the meters may be above or below the frequencies that you can hear. Just because you cannot hear any sound does not mean that it won't negatively interact with the audio you can hear. If you see or hear a marked increase in the noise, you can be sure that it's being added to whatever you're doing. This is not good.

If raising the console's master gain control does not result in noise, continue your search by raising each channel independently to the previously marked operating level. If you detect noise on any channel, it's likely that you have a problem with the channel or with the audio source feeding it. If unpatching or disconnecting the source eliminates the noise, check the source or the connection. If the noise is constant with and without the source connected, the problem is most likely in that console channel.

Gain Balancing

Before you determine that a noise you hear is due to a malfunction, check the output level of the device feeding the console and the input attenuator of the console, if it has one. You may find that one stage is turned up too far to compensate for the other being too low. Try rebalancing the controls so that each is operating at a more moderate level. As a rule of thumb, most circuits are designed to operate best when turned up to three-quarters of their full range. Exceeding that amount normally results in increased noise, which is audible as hiss.

External Electronic Noise

In addition to acoustic and internal electronic noise, external electronic noise also can compromise your production. Even though external electronic noise can be heard in the system, its source is outside the system.

Radio frequency (RF) noise from radio and television transmitters, data transmitters, and any device that uses RF as part of its operating system can be inadvertently "received" by parts of your system. In the case of audio-modulated RF, the audio portion may be "demodulated" by your system, allowing you to hear the audio content of the external source.

Because the sources are sometimes intermittent, such as the transmission of two-way radio communications, it can be difficult to pinpoint the part of your system that is receiving the unwanted RF. More constant external RF noise sources, such as 60-Hz energy radiated from nearby power sources and fluorescent light fixtures, can also be received and are easier to isolate. This noise is often received by or before a high-gain amplification stage in your system, such as a mike preamp. The unwanted noise combines with the relatively low signal level of the system. The high-gain amp stage then amplifies the source audio as well as the external noise.

Unbalanced circuits, which offer less shielding, generally are more susceptible to external noise. Chapter 3 discusses the differences between balanced and unbalanced audio. Loose or bad connections anywhere in the system may also be the culprit. No matter how well the system is shielded, the proximity of the noise source to the system or the intensity of the noise source sometimes makes eliminating the noise impossible. When this happens, you have to move either the source or the affected part of the system.

If no appreciable noise is detected on any individual channel, the next step is to raise all faders or pots, including the master, to their marked positions. Again, note any audible noise or any increase in output meters. Grounding and other problems that are not measurable on individual channels can sometimes be detected when the channels are combined.

Some CD players, turntables, and tape machines have mute circuits built into their output stages. When no audio is output from them, the mute turns off all audio, including any noise that they may generate. This can fool you into thinking your system is quieter than it really is. If you suspect that this is the case, listen between CD or record cuts, or play a roll of virgin tape, that has never been recorded or erased. Any noise you detect will be the result of problems with the source or bad connections at the console.

CONCLUSION

The best producers are those who have not simply mastered an awareness of the audible frequency spectrum, but are also extremely aware of music, talent performance, microphone choices and techniques, equipment interfaces, the symptoms of technical problems, a lengthy menu of audio processing possibilities, and a good measure of human psychology. The ability not to take yourself too seriously is also a bonus.

Raising yourself to successive levels of proficiency takes time, practice, and a lot of listening. Your ability to improve as a producer is directly tied to your ability to listen. If you've had some experience with audio production already, you can probably tell the difference between your earlier work and your present work. This learning process never stops. Some people develop the ability to mentally process audio easily and quickly. Others require more time.

Getting Connected

3

The more you know about the technical aspects of a particular system or piece of equipment, the greater your spectrum of possibilities. For a producer, however, there is a point of diminishing return. Past this point, a producer tends to focus too much on the equipment and not enough on the sound and the content. New producers, in particular, have a tendency to become more fascinated with the equipment and what it can do than with the outcome of the project. It takes a while to learn that it's as important to know when not to use effects as it is to know when to use them.

The producer of today is more concerned with how to operate a device and with what it can do than with the *how* or *why* of its design. It is no longer as necessary to know every internal detail of a piece of equipment to use it well. In fact, due to the speed at which technology is advancing and the number of new audio products on the market, complete comprehension is practically impossible.

It is not by coincidence that this phenomenon parallels the increase in the use of computers in today's workplace. Most of today's audio production gear has some sort of computer-based operating system on board. Audio design engineers and software developers or programmers are often one and the same. Like it or not, producers are becoming computer operators.

The producer faces another problem with equipment manuals that are written more from a design or technical perspective than from a user's point of view. The use of too much technical jargon and the failure of the manuals to give practical information and examples create a needlessly steep learning curve for anyone who lacks a technical background.

Another factor that works against understanding is that equipment is often developed and manufactured in countries where different languages are spoken. Confusion and misunderstanding created by bad translations and editing occur much too frequently.

Producers must make a continuing commitment to adapt to new technology. Part of that commitment is accepting the fact that new ideas take time to absorb. You must also learn to make judgments as to which of the new designs will improve the quality of your work and will allow you to do that work more quickly. You will also have to be prepared to reinvent yourself periodically as the benefits of new pieces of equipment cause you to change your approach to any or all of the basic elements that you have already learned.

Take EQ, for example. If the music from an analog tape recording is fed through an analog equalizer set to boost all of the frequencies above 12 kHz, the resulting audio sounds brighter, but the circuit noise and tape hiss are also increased. In an all-digital system, however, tape hiss and circuit noise barely exist. The lack of tape hiss and circuit noise are obvious benefits. It's up to you as a producer to recognize that the change offers the possibility of using more high-frequency EQ than in the past.

GETTING PAST THE JARGON

Your best protection is a thorough understanding of all of the ways that pieces of equipment can be connected in the studio. This starts by knowing as much as possible about the various inputs and outputs of the equipment. Because of the increasing number of options, both in the number of audio-processing devices and where and how they may be used in a production studio audio chain, it's easy to get sidetracked or bogged down by the technology.

Although it's impossible to know everything, a few basic electronic concepts, such as where various pieces of processing gear fit in a production audio chain and why, are of prime concern to the producer. Without an understanding of balanced and unbalanced lines, input and output levels, and impedances, you can make mistakes that compromise the work being done or damage the equipment.

The truth is that mistakes are not uncommon. You must be constantly vigilant for any changes in the quality of the audio. If the cable you grabbed to connect a new or loaned piece of equipment is not wired correctly for that piece of equipment, you had better know about it before it wrecks an entire day's work.

IMPEDANCE

Impedance is defined as the resistance to the flow of alternating current (AC), which is what analog audio is. The basic plumbing of an audio circuit consists of resistors, inductors (coils), and capacitors, or condensers. Most audio circuits conduct a variety of AC and direct current (DC) currents and voltages.

Resistance

Resistors pass both AC and DC. A hot plate and an incandescent light bulb are good examples of pure resistance at work. The 60-Hz AC (house current) moving back and forth within the wires of the hot plate create friction that heats the wires and your coffee. The thin filament within a light bulb responds in the same way, but because there is no oxygen in the vacuum created inside the bulb, the filament can reach white-hot temperatures and creates light without burning up.

Inductance

Coils of wire, such as transformers, pass DC easily, but resist the flow of AC. The magnetic fields generated around a conductor carrying DC do not change as long as the DC flow remains constant. With AC, the direction of the current is always changing. As these changes occur, the magnetic fields generated by the current flowing in the conductors constantly expand and collapse each time the current changes direction. Because the wires of a coil are wound so closely together, the collapsing and expanding fields from adjacent wires interact with each other, presenting an impedance to the AC. This type of impedance is called *inductive reactance*.

Capacitance

Capacitors consist of two surfaces separated by a nonconductive material called a *dielectric*. Because its two surfaces are not directly connected, a capacitor will not conduct DC. It will pass an AC signal, however, even though the plates do not touch and there is no direct connection. The capacitor is designed to take advantage of the changing directions of AC. As the current swings back and forth, the charges on the plates of the capacitor also alternate. It is the change in charge between the plates caused by the AC that passes through the capacitor, not the actual current itself. In addition to their ability to block the flow of DC, capacitors are designed to present greater impedances to some types of AC than to others. This capacitive reactance is used in circuit design to control which AC frequencies may pass through a circuit.

Rules of Impedance

Impedance, then, is the combination of pure resistance, inductive reactance, and capacitive reactance. Each input and output of a piece of equipment has a measurable impedance.

As a rule of thumb, when connecting the output of one device to the input of another, the best results are obtained if the output impedance of the device is the same as or lower than the input impedance of the device you are plugging it into. Look for the input and output impedance figures of a device in its operating manual. Impedance is always measured in ohms.

Although you may be able to get audio from one piece of equipment to another without having their impedances match, you may not care for the sound of the audio. Typically, the side effects of impedance mismatch are levels that are too low or too high for the input of the piece you are attempting to plug into. This is known as a *nonlinear frequency response*. The high or low frequencies are the first to be attenuated by this sort of bad connection. The negative effects are sometimes very noticeable, and sometimes not. It all depends on how bad the mismatch is and how forgiving the equipment is.

If, in the past, you have found impedance requirements confusing and elusive, consider taking a simpler, more visual approach to the concept. In Figure 3.1, the large water pipe represents the low-impedance output of one device. The small pipe, which is fed by the large pipe, represents a high-impedance input. The impedance, or restriction to flow, in the large pipe is many times less than that of the small pipe. Although the small pipe can't handle the volume of the large pipe, it will remain full and continue to carry water.

Now imagine the flow if the connection is reversed (Figure 3.2). If the output of the small pipe is fed to the large pipe, which has a relatively lower impedance, a solid, consistent flow becomes impossible.

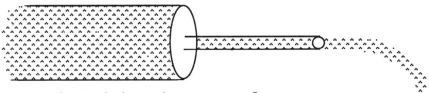

Low impedance to high impedance—constant flow

Figure 3.1 Constant flow due to acceptable impedance match. A large pipe with a relatively low restriction (impedance) can feed a smaller pipe with a higher restriction (impedance) with a smooth, continuous feed.

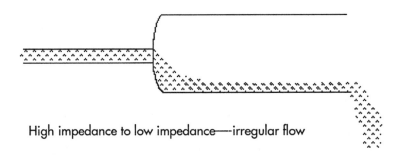

High impedance to low impedance—irregular flow

Figure 3.2 Irregular flow due to impedance mismatch. A small pipe with a relatively high restriction (impedance) cannot smoothly feed a larger pipe.

Figure 3.3 shows two devices, each with stereo inputs and outputs. Device A on the left is a high-impedance device with 100-kiloohm inputs and 10-kiloohm outputs. Device B is a low-impedance device, with 100-ohm inputs and 600-ohm outputs. Feeding the lower 600-ohm output of device B into the higher 100-kiloohm input of device A will make a good connection because 600 ohms is less than 100 kiloohms. Feeding the 10-kiloohm output of device A into the 100-ohm input of device B is not a good connection because 10 kiloohms is more than 100 ohms.

Most equipment is made with inputs that are relatively higher in impedance than the outputs. This makes connecting similar devices easy. Unfortunately, it is not at all unusual to find a mix of both high- and low-impedance devices in the same system.

A wide variety of impedance-matching transformers and match boxes that compensate for bad matches are available. Transformers are passive, requiring no power. They comprise two sets of wire windings wrapped

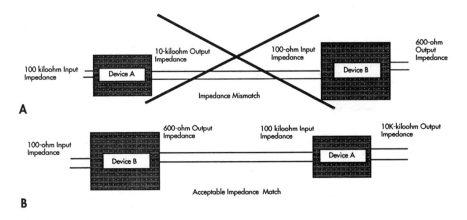

Figure 3.3 Two ways to connect a high-impedance device (device A) with a low-impedance device (device B). (A) Feeding the output of device A into the input of device B results in impedance mismatch. (B) The reverse arrangement results in an acceptable impedance match.

around a common core. The windings are designed to provide each of the devices with the correct impedance. The signal is passed through the transformer by induction rather than by direct connection. The AC audio output of the first device feeds the input, or primary side, of the transformer. The AC of the signal on the primary side creates magnetic fields that are picked up by the windings of the output, or secondary side, of the transformer, which has been designed to provide the proper input impedance to the device it is plugged into.

Transformers are also used to boost the level of a signal or to drop it so that it more closely conforms to the level range of the device before it. With a step-up transformer, the secondary side has more windings than the primary side. With a step-down transformer, the secondary side has fewer windings than the primary side.

Although transformers have their uses, even well-designed ones take their toll on the audio that passes through them. In theory, a 1:1 isolation transformer with identical primary and secondary windings is virtually invisible to the AC that passes through it. The same level of audio applied to the primary side is theoretically available on the secondary side. In practice, however, some of the electrical energy is lost in the inductive transfer inside the transformer. Cheaper, poorly designed models may further compromise the quality of the audio that passes through them by restricting the bandwidth of the audio, much the same way an impedance mismatch does.

Match boxes use transformers and other circuit components to provide correct impedances and levels to improve the quality of the connection of devices. Unlike transformers, match boxes use powered circuits, which use additional electrical energy to overcome losses. It comes back to the concept of knowing your options, and making judgments.

OPERATING LEVELS

In addition to impedance differences, audio inputs and outputs can also have level differences. These are often referred to as *semipro* and *pro* levels, with pro levels being higher than semipro levels. Because semipro levels are normally about 10 dB lower than pro levels, a 0-dB feed from a semipro output will normally drive a piece of pro gear to only −10 dB. When pro gear feeds semipro gear, the output of the pro gear or the input of the semipro gear must be reduced, or padded, to keep the input of the semipro gear from being overdriven.

A major cause of confusion is that the meters on both semipro and pro equipment look and are read exactly the same. This is because meters are calibrated to read the dynamic range of the audio that passes through them, relative to their own operating levels.

Although more boxes are being designed with inputs and outputs that can be switched to either semipro or pro, there are still quite a few circuits that are not as flexible.

Match boxes can be used to boost and balance the −10-dB level of the semipro gear up to the 0-dB or +4-dB level at which most pro gear operates. When considering these boxes, try to use only the highest-quality circuits to prevent signal deterioration.

HEADROOM

As you are probably aware, most circuits have some sort of metering: volume unit (VU) meters, peak flashers, light-emitting diode (LED) meters, or plasma displays. Different metering circuits, even though they may appear similar, may not be reading the same thing. Meters with moving parts are the least responsive of all. Because they are mechanical, they are physically incapable of showing transients, or very brief, very loud passages of audio.

This explains why peak flashers, which are more sensitive to the peaks of the audio that pass through the circuit, often flash before the needle of a mechanical meter reaches the red.

Headroom is the amount of signal over 0 dB, the normal operating level, that a circuit will pass before it distorts. Headroom varies widely from circuit to circuit. This numbers game can be very misleading. Two consoles that claim the same signal-to-noise (S/N) ratio, a figure representing the level above the noise floor at which they operate, may seem equal to each other. A closer look at their headroom may show that the extra headroom of one console allows it to pass considerably more audio before distortion than the other. The safety that extra headroom provides is particularly attractive when you are working with audio that has lots of unexpected peaks.

DYNAMIC RANGE

Dynamic range is simply the spectrum of sound, from loudest to softest. Each device that originates, passes, or receives audio has its own dynamic range. As shown in Figure 3.4, if the audio passing through a circuit is be-

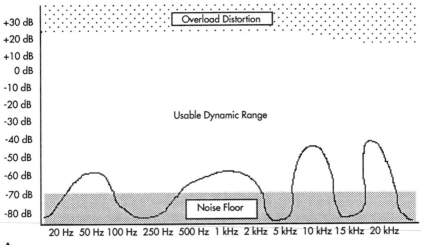

A

Figure 3.4   (A) A signal that falls below the noise floor. Noisy signal results. (B) A signal that exceeds the headroom of the circuit. Distorted signal results. (C) A normal signal above the noise floor, but below distortion. This is the correct level.

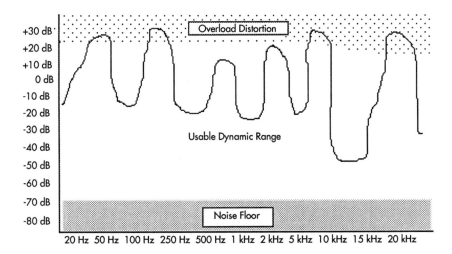

B

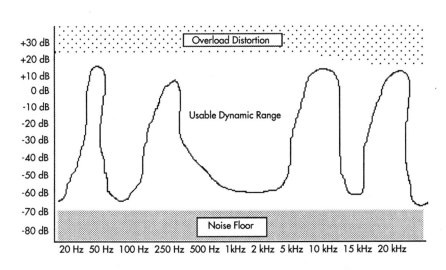

C

low or above its recommended dynamic range, one of two things happen: Low levels become lost in tape hiss and circuit noise, or levels that exceed the dynamic range become distorted. In film and video, the comparable phenomenon occurs when too much light causes overexposure and too little light causes underexposure.

Failure to keep operating levels above the minimum recommended dynamic range can also cause other problems, even in the digital domain. When analog audio is converted to 16-bit linear pulse code modulation (PCM) digital audio, the analog-to-digital (A/D) conversion applies all 16 bits of resolution to the loudest passages. As the level of the signal decreases, fewer bits are used in the conversion. When the level drops below a point, there are simply not enough digital bits to define the converted audio accurately.

Film speed and light can be used to make a comparison. A camera system set for 100 ASA/ISO film in bright light would automatically switch to 125, 200, and 400 ASA/ISO as the light decreased. At the lower light levels, the faster film speed offers reduced definition, or grainy prints.

BALANCED AND UNBALANCED AUDIO

In Figure 3.5, the standard balanced pro-level audio leaves the output circuit via a three-conductor connection called a *Cannon* or *XLR* connector. Conductor 1 (pin 1) is normally a shield wire that carries no signal and connects to the ground. Conductors 2 and 3, connected to pin 2 and 3, respectively, normally carry the alternating audio signal.

Manufacturers differ on whether pin 2 or pin 3 is hot. Look for this kind of information in the equipment operating manual. If pins 2 and 3 are inadvertently reversed, the polarity of the audio will be reversed by 180 degrees. Simple devices like light bulbs and electric clocks are not affected if you re-

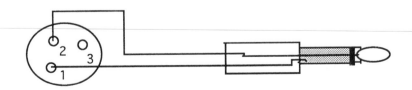

Figure 3.5 One of the acceptable ways to unbalance a balanced line.

verse the polarity of their power by changing the way you plug them into the wall. With audio, however, it is important to maintain consistent polarity throughout the entire system. The audible effects of having one of two stereo channels out of phase with the other may not be readily apparent. However, if the two channels are combined to monaural, some of the audio may cancel out (Figure 3.6).

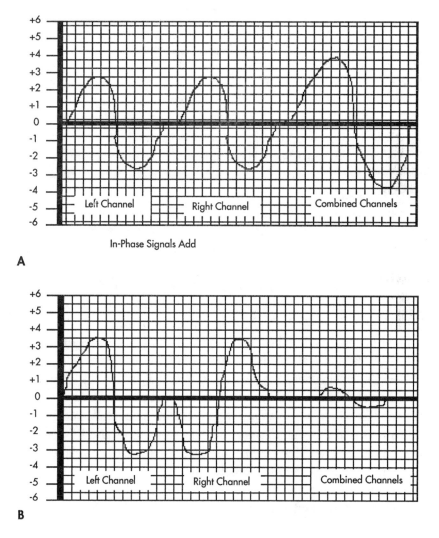

Figure 3.6 The effects of cancellation. In (A), in-phase audio from both the left and right channels increase the total output when mixed together. In (B), out-of-phase audio from both channels decreases the total output.

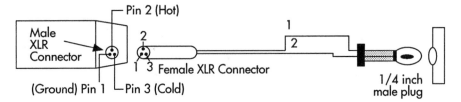

Figure 3.7 Male XLR output jack from a chassis with pins 1, 2, and 3. Pin 1 is ground, pin 2 is hot (positive), and pin 3 is cold (negative). The connecting cable has a female XLR connector on one end and a ¼-inch TS plug on the other. The ground wire from pin 1 of the female XLR is wired to the barrel of the ¼-inch jack plug. The hot pin 2 audio is connected to the tip of the ¼-inch TS plug. The input of the unbalanced device is shown as a two-conductor ¼-inch jack mounted on a chassis.

Even without a match box, depending on the specific configurations and designs, driving semipro gear with pro levels can be very easily done. Since the level and impedance requirements of a device are usually explained in its operating manual, it's always better to check there before doing any experimenting.

A specially wired connecting cable is needed to unbalance the output of a balanced device so that it can feed the input of an unbalanced device. As shown in Figure 3.7, the cable must connect the ground from XLR terminal 1 to the barrel or ground of a ¼-inch phono plug and must connect 2 or 3 (whichever is designated the hot wire) from the XLR to the tip, or high side, of the ¼-inch phono plug. This unbalances the three-conductor pro gear's output so that it can drive the two-conductor semipro gear.

Although you may find the audio quality of unbalanced lines more than acceptable, unbalanced audio lacks the separate conductive metallic film or braided strand shield offered by a balanced line. It also runs at a lower level because it uses only one of the conductors carrying the audio. Although an unbalanced line is more susceptible to electronic noise caused by bad ground systems and by interference from nearby power and RF sources, such as radio and television stations or citizens band and taxi radios, you may find that it works just fine for your needs.

As illustrated in Figure 3.8, because unbalanced audio in a semipro configuration normally runs at a lower level than pro, any noise added to the signal will be proportionally greater and more apparent than the same amount of noise added to a pro-level signal.

Note that the presence of XLRs or ¼-inch phono connections does not necessarily confirm whether a line is balanced or unbalanced. A few devices

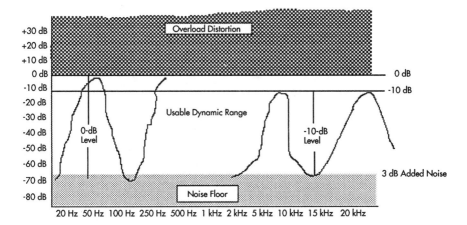

Figure 3.8 Identical side-by-side waveforms, one at –10 dB and one at 0 dB, each with the 3 dB of noise superimposed to show the relative differences.

The waveforms are identical except for their levels. The higher the level of the waveform on the left indicates 10 dB more usable dynamic range than the waveform on the right.

use two conductors of a three-conductor Cannon connector, but they are unbalanced. In addition, some balanced devices use three-conductor ¼-inch tip–ring–sleeve (TRS) connectors. Check the manual for this information.

Unlike the fairly standardized types of audio inputs and outputs of the last 30 years, recent advances in solid-state circuit design may require special attention. Check the manual for the best ways to hook up the device. Special considerations are usually mentioned in the manual.

4

Understanding Audio Signal Flow

The plain truth is that many people who regularly work with audio on a particular system know only enough to do very specific jobs. They may be competent operators, but they are not producers. Among other things, becoming a producer means learning as much as possible about every piece of equipment in the system on which you work. Just as the producer develops the ability to change her listening perspective from the smallest detail to the total sound of a production, it is also necessary to know the workings of each piece of the system as well as the many ways those pieces can be connected.

No matter how imposing and complicated any given studio system is, all systems serve the basic function of sending an electronic audio signal from one place to another. That signal may pass through a variety of mixers, effects devices, amps, and monitors, but all signals—analog or digital—follow a predetermined mappable route. When you understand how the signal is routed, you have control over the system.

Because even systems that use the same components can be configured very differently, it's better to start by thinking about signal routing in more general terms. Where is the signal coming from? Where is it going? What piece of equipment does it go through first? Second? Where does the signal end up?

It may help to draw a simple diagram of your system. Visualize the pathways that audio can follow in the system. If the system is complicated, the

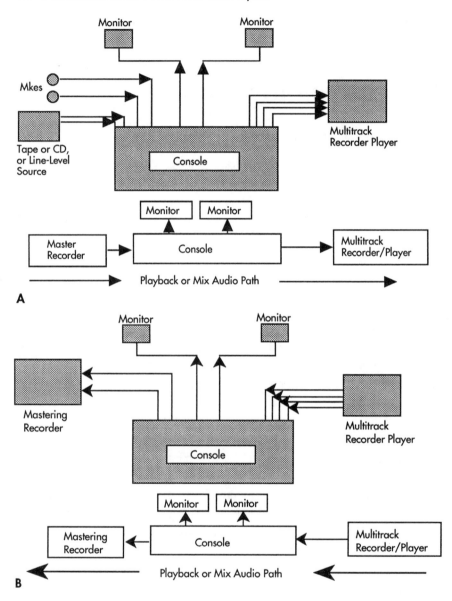

Figure 4.1 Sample signal pathways. (A) A simple radio production setup. (B)A multitrack music setup.

signal may take more than one path at a time. Make separate diagrams for each configuration. Figure 4.1 shows examples of such diagrams.

Note the possible pathways. Even if you have been working with the system for a while and feel you know it well, it is a good idea to brush up on all of the inputs, outputs, and other connecting possibilities from time to time. This kind of creative exercise has two advantages. First, the more you know about your system and the more comfortable you are with it, the more options you have. Second, once you are aware of your options, you have the control you need to handle unusual client requests, emergencies, and creative brainstorms. The very act of developing different routing possibilities is good practice in itself. As in most crafts that require equal parts of technical, aesthetic, and operating abilities, the best results occur when all parts are employed. This means that you must be as creative with how you use your system as you are with what you produce.

For example, a client's request for a seemingly simple change in the structure of a piece will force you into uncharted territory. You may have run out of tracks or microphone inputs or have found yourself stymied by any number of developments. The true professional never lets the problems show. The better you know your system, the better your chance of surviving these experiences successfully.

Suppose one side of the main stereo line amp in the console stops working just as you are ready to record a stereo master to a two-track reel-to-reel tape machine. The console may have another pair of alternate line outputs or other outputs of the correct impedance and level to feed the two-track.

The last chapter described how impedances and levels determine how to connect different pieces of equipment. This works well as long as the system is fairly simple, such as the one shown in Figure 4.2. In this system, there is no need to change any of the connections, but suppose that, in addition to the two-track reel-to-reel recorder, the studio gets a DAT (digital audio tape) recorder for mastering productions. If the console only has one stereo output, the producer must change the console's output cables to whichever machine is needed to make the master.

As more pieces of equipment are added to the system, each with a number of additional inputs and outputs, it becomes convenient to have a point in the studio to which all equipment is connected. The simplest device for this kind of connection is the patch bay.

PATCH BAYS

The patch bay brings together in one place all the inputs and outputs of the console and the other parts of the system. Instead of each piece being directly connected to each other, as in Figure 4.2, the cables are attached to

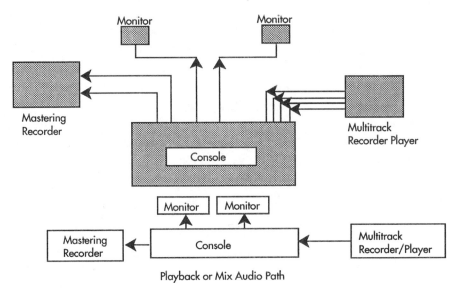

Figure 4.2 Simple diagram of studio system, showing equipment connections.

labeled jacks in the patch bay. Patch cables are used to connect the outputs of one piece of equipment to the input of another, much the same way that a telephone switchboard lets the operator connect an incoming call to the right telephone. Each jack in a patch bay connects to a cable that is attached to a piece of equipment in the studio. This greatly simplifies the procedure of connecting various pieces of equipment to the console and to each other during different stages of production. For example, the outputs of the console can be connected to, or "jacked into," either the reel-to-reel mastering recorder or the DAT mastering recorder. The output of the recorder being used is then patched into the monitor amp, allowing you to hear the mix in the monitors. Later in this chapter, more expensive consoles with features that eliminate the need for this kind of patching are described.

Although some standards in patch bay configuration do exist, the best rule is never to assume that a patch bay with which you are familiar bears any resemblance to one you have never used. The labeling of the patch bay should give you an indication of what inputs and outputs are available, but due to limited space, the labeling may be quite cryptic. The generic term patch *point* is used to identify an input or output that can be connected to a *patch* bay. Most mixing consoles have many patch points, allowing the producer to route audio via the patch bay to other places in the console or to other pieces of equipment in the studio.

Normal, Double-Breaking, and Half-Breaking Jacks

Standard patch bays connect the console and the equipment. *"Normalled" patch bays* use special jacks with extra contacts to form a connection when no patch cable is in the jack. The inputs and outputs of each part of a normalled system are connected by jumper wires connected to the appropriate outputs and inputs of each piece in the system. This allows a particular collection of gear to be connected through the patch bay without the use of patch cables.

If the normalled jack is a double-breaking jack, inserting a patch cord will break the normalled connection. If the jack has been wired to the output of a tape recorder, for example, the audio from the recorder will not pass through the broken normalled connection. Instead, it will be available at the other end of the patch cord and can be plugged into an appropriate input. In contrast, a half-breaking normalled jack allows the audio to continue to pass through the normalled connection, but it also makes the signal available at the other end of the patch cord. Some patch bays combine the features of both normalled and standard connections. With this option, only the parts of the system that are most frequently configured the same way are "normalled." Other parts of the system can then be patched in as needed. (See Figure 4.3.)

In a typical radio station production studio, the console outputs may be normalled to the input of the cart machine that is used to record carts. The console's monitor outputs are sometimes normalled to the power amp that feeds the studio monitors. This allows the monitors to function normally without being patched in, and it allows other line-level outputs, such as individual tape machines, to be patched directly to the monitor amp. Being able to patch any line-level device directly into the power amp and monitors is convenient when you're trying to isolate an audio problem. (See "Listening for Noise" in Chapter 2 for more details.) Regular and normalled patch bays look the same from the front.

Balanced and Unbalanced Patch Bays

Some patch bays are balanced, using three conductors: a "hot," or positive, wire and a "cold," or negative, wire for the audio, and a separate shield, or ground. The plugs on these patch cables are TRS. The standard wiring for this plug is hot to tip, cold to ring, and ground to sleeve.

Other patch bays are unbalanced, using only two conductors in a tip–

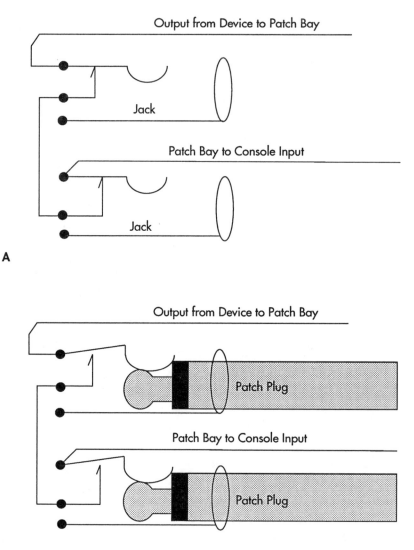

Output from Device to Patch Bay

Jack

Patch Bay to Console Input

Jack

A

Output from Device to Patch Bay

Patch Plug

Patch Bay to Console Input

Patch Plug

Single or Half-Breaking Configuration

B

Figure 4.3 In (A) the connections and moving parts of a full or dual-breaking patch panel jack is shown. The first illustration shows a closed jack without a patch plug inserted. The lower illustration shows how the insertion of a patch plug opens the normalled contact and routes the signal to the tip of the inserted plug. (B) Shows the same action for a single or half-breaking jack.

sleeve (TS) configuration. The standard wiring for this plug is hot to tip and ground to sleeve. In unbalanced patch bays, the ground or shield serves as the cold side of the unbalanced circuit.

Interchanging TRS and TS patch cords is not a good idea. Although you might be lucky, you also run the risk of short-circuiting the inputs or outputs of a piece of equipment. Although much of today's equipment is built to withstand a direct short and some is designed to accept either TRS or TS, some older gear is not. Shorting equipment that is not protected can cause distortion, noise, or a compromised frequency response, or it can damage the circuit.

PHASE REVERSAL AND PHASE CANCELLATION

If the hot and cold connections at the patch bay or in any cable between the outputs and inputs of two pieces of equipment are reversed, the problem may go unnoticed until you attempt to mix that audio with correctly connected inputs or outputs. Even if you are doing a simple stereo mix, the problem may not be immediately apparent. The real test is to listen to the mix in mono. When you listen to a stereo mix in mono, the stereo separation disappears, and the sound seems to come from a point directly between the monitors. However, other than the loss of the stereo image, you should hear no difference in the level or frequency response of the individual elements in the mix. This is called *checking for mono compatibility*. If you do hear a difference, you probably have a few wires reversed somewhere.

The importance of this concept becomes easily apparent when you look at what happens to a sound as it travels through a system. Suppose that, as in Figure 4.4A, we use a system that comprises a mike, a mike preamp, a mixer, a power amp, and a speaker to amplify the sound of a drum being struck. When the drumstick hits the drum, a sound wave is created that pushes the moving part of the mike in the same direction. If there are no phase reversals in the system, the speaker will move in the same direction when the sounds reaches it. This system is in phase.

In Figure 4.4B the wires have been reversed at one stage, causing an out-of-phase situation from that point on. Now, instead of pushing outward when the drum is struck, the speaker pulls back. Although at first listen the sound from the speaker in Figure 4.4A may be indistinguishable from that

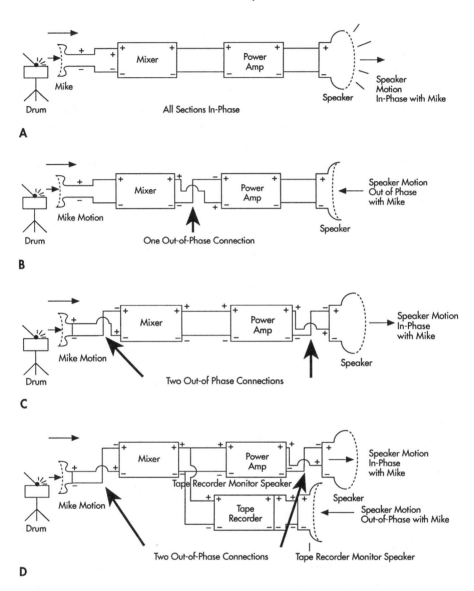

Figure 4.4 (A) An in-phase system. The drum is struck by the drumstick, distending the drumhead and pushing the microphone element in the same direction. Positive and negative signal wires are connected to all parts of the system in phase. The speaker element distends identically to the drum head. (B) An out-of-phase system. One connection is reversed, causing the output of the speaker to be out of phase with the mike. (C) An in-phase system. Two connections are reversed, which apparently corrects the problem. (D) The system shown in Figure 4.4D with an in-phase circuit attached between phase reversals. An out-of-phase condition results at the tape recorder speaker.

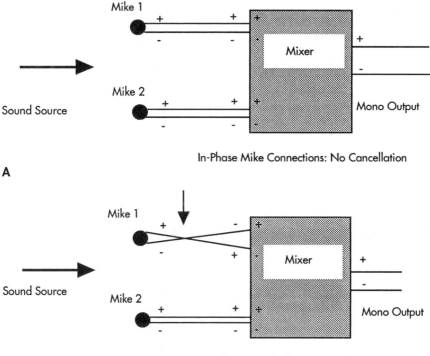

A

B

Figure 4.5 (A) Two mikes at the same source with all lines in phase and combined before the power amp. (B) Two mikes at the same source with one circuit out of phase. The signals combine and cancel each other.

in Figure 4.4B, engineers and producers with a well-developed listening ability can tell the difference.

In Figure 4.4C it appears that two wrongs make a right. The out-of-phase condition created by the first switch is corrected by the second, so that the drum, mike, and speaker are all moving in the same direction. However, if a circuit with no phase reversal is connected between the two phase reversals, its output will be out of phase, as shown in Figure 4.4D.

Now consider Figure 4.5A, which shows what would happen if two different mikes and preamps were combined into one signal before they reached the speaker. As long as both circuits remain in phase, the motion of the drumhead being stuck is repeated exactly by the movement of the speaker. However, if the circuits are out of phase with each other, as in Figure 4.5B, the signal that pushes the speaker cone out will cancel the signal

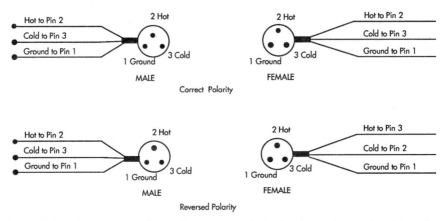

Figure 4.6   The top figure shows that correct polarity has been observed and that the connection is in phase. In the lower figure, the reversed polarity between pin 2 and pin 3 on the right side create an out-of-phase connection.

that pulls the speaker in, resulting in no sound being heard. This effect is called *phase cancellation*.

CABLE POLARITY

Being aware of cable polarity is very important because there is no international polarity standard for balanced audio circuits. Some balanced circuits use conductor 2 as the high, or positive, conductor; others use conductor 3. To make matters worse, in some devices, the polarities at the inputs are opposite those at the outputs, so although conductor 2 is high at the input, conductor 3 is high at the output. As Figure 4.5 indicates, this results in situations in which the audio may or may not be in phase, depending on which pieces of equipment are connected and in what order they are connected.

It is a good idea to keep track of the polarity differences of different pieces of equipment when making up cables that connect to the back of the balanced patch bay. Some studios keep a numbered log of the polarity of each device's cable. Another way to keep track is to make labels for each cable. Attach the labels to the cable just before the connectors that plug into the particular device. A simple label marked "2 Hot" or "3 Hot" lets you quickly identify the polarity of the connection. If at some point in the future you decide to change devices, you will immediately know by reading the label if the polarity of the connection is correct for the piece of gear you wish to connect.

In addition to using a patch bay, studios that cater to clients who bring in their own equipment often have a separate interface panel with several different types of jacks, such as XLRs, RCA cinch jacks, and ¼-inch TRS jacks. These jacks are connected to the jacks in the patch bay. The interface panel is convenient because it allows any equipment brought into the studio to be connected to and integrated with the patch bay. Instead of making and mounting a panel, some studios connect several balanced or unbalanced cables to unused jacks in the patch bay and put standard XLR or ¼-inch connectors on the other end. The cables are made long enough so that they will reach places in the studio where equipment can be temporarily installed.

DISTRIBUTION AMPS

The more standardized and "normal" the operation, the less the need for a patch bay. However, you may still need to route audio from one source to several other destinations simultaneously at some point. If your console doesn't have enough outputs, a distribution amplifier (DA) can be used. It takes one input—mono, stereo, or both, depending on the design—and distributes the signal to a number of outputs simultaneously. For example, if a stereo console has only one stereo output, it can be fed to a 1×3 stereo DA, which can feed stereo to three separate devices—for example, a tape recorder, a headphone amp, and a monitor amp, as shown in Figure 4.7.

If your console has no effect or auxiliary sends (outputs) and the reverb you would like to use cannot mix effected signal with dry signal before its outputs, you can use one of the DA outputs to drive the effects box. As Figure 4.7 shows, the output of the effects box can then be patched into another console input and mixed in with the dry signal.

The design of the DA dictates its versatility. Different manufacturers sometimes design their DAs differently. A 1×3 DA usually means that one mono input is fed to three mono outputs. A 2×6 DA usually means that one stereo input is distributed to three stereo outputs. A 1×3 stereo DA usually means that one stereo input is fed to three stereo outputs. Some DAs are designed to allow a stereo signal to be combined to mono at any or all of its outputs. Your understanding of the possibilities and the expected uses dictates which DA will do the best job.

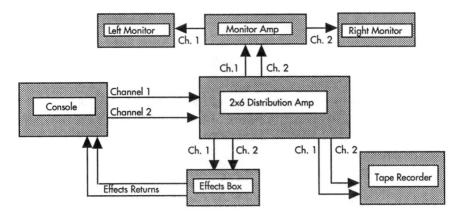

Figure 4.7 A console feeding a 2 × 6 stereo distribution amp. One distribution amp output feeds a tape recorder. Another feeds an effects box, and the third feeds a monitor amp. The output of the effects box is fed back to the console.

AUDIO SWITCHERS

Audio switchers are like electronic patch bays. Like patch bays, they connect different pieces of a system. Unlike patch bays, however, switchers use a matrix of internal connections, instead of jacks and patch cords. The audio inputs and outputs of different pieces of the system are directly connected to the rear of the switcher. The routing occurs internally as a result of manual front panel operation, remote control via a computer or a simple remote control panel, or program change information from a musical instrument digital interface (MIDI) device. (More on MIDI in Chapter 6.) With most switchers, configurations can vary from having any input drive one or all of the outputs to having each input feed a separate output.

Audio switchers vary in design, but the good ones share some features, including adjustable inputs or outputs and a nonvolatile memory. The memory is extremely useful if you are using a system in a number of predetermined ways. Once you have achieved the routing for a specific job, a number is assigned to that routing configuration. The next time you need that setup, all you have to do is hit the recall button, and you're ready.

CONSOLES

Because each console is different and even identical models can be configured in completely different ways, knowing as much as possible about the specific configuration of the console with which you are working is the best way to understand its options. The most logical way to approach this task is to start from the front, or input, and work to the back, or output.

The front of most console channels is set up to receive mike-level or line-level sources. Mikes connected to the mike inputs of a console are routed through the channel's mike preamp. This preamp brings the relatively low mike level of –20 dB to –45 dB up to line level. Line-level sources, which usually operate from –10 dB to +4 dB, don't need preamplification. Even so, line-level inputs often have an input trim control that varies the amount of signal that feeds the input strip. Balancing the output of the device feeding the console with the input trimmer for the appropriate amount of gain with the least amount of noise is discussed under "Gain Balancing" in Chapter 2.

There are two important things to remember here: First, the trimmer should be used to set the level into the console so that the source does not overdrive or distort the input. Second, the trimmer should allow enough level to the input stage so that extra gain is not needed to bring the source up to a good mixing level. If the input level is too low, attempts at increasing the level will result in increased and unwanted circuit noise.

Switchable Inputs

Many console inputs are switchable. This means that you can plug several different sources into the input and switch to the one you need at the moment. For example, you can have a mike and a tape machine connected to the same input channel. This configuration requires that you switch back and forth, depending on which source you need.

Although a few consoles have inputs that are designed to route more than one source at a time, they restrict the amount of independent control you have over each source. Any changes you make are made to both sources.

Simultaneous Mike/Line Inputs

Some consoles are designed with the ability to pass both mike-level and line-level signals through the same input at the same time. Typically, the mike and line each have their own input level controls, or trim controls. Trim controls allow the two signals to be mixed to the proper levels as they enter the mixer.

Mixing can be done even if only one side of the input has a trim control.

If only the mike input can be trimmed, you may still be able to mix the two sources by varying the mike gain against the nontrimmable line input. If too much adjustment of the mike trim causes unwanted circuit noise or distortion, you may still be able to mix the two inputs if your line-level source has a variable-output control. If it does, try setting the mike trim for the best results and adjusting the output of the line-level device for the proper mix.

Inserts

Inserts are simply intentional breaks in the audio line in which you can insert additional pieces of equipment. If, for example, the console has very little EQ, you might choose to insert an equalizer to get the sound you need. Inserts are usually brought out to the patch bay, making it easier for other pieces of gear to be placed in the audio chain.

Foldback Circuits, Auxiliary Sends, and Effects Sends

Foldback circuits, auxiliary (AUX) sends, and effects sends all operate on the same principle. They allow you to divert a variable amount of audio from a source without affecting the original signal. This diverted audio is called a *send* because it is sent from the original audio for some other purpose. Depending on the specific console and how you use it, foldback circuits, AUX sends, and effects sends may be used interchangeably. In the simplest configuration, the send is often routed from the console to the input of an effects box such as a reverb. The output of the reverb is then routed back to the console via an effects return circuit.

Depending on the console design, the effects return circuit may have a gain control that allows you to adjust the amount of returning reverbed audio. If it doesn't, you have two options: The effects device may have an output gain control by which you can vary the amount of reverbed signal returning to the console, or you can route the output of the reverb to an unused input on the console and mix in the required amount.

In addition to a system of several effects sends and returns, more sophisticated consoles may offer AUX sends. These AUX sends can be combined to make additional separate mixes from the master mix. In live music applications, AUX sends may be used to create a mix that is fed to the stage monitors, allowing the musicians to hear what they are playing or singing.

In the studio, AUX sends are used to create a separate headphone mix so that the performers can hear whatever parts of the recording are most helpful while overdubbing.

Whether they are mono, stereo, with EQ, or without it, foldback circuits, AUX sends, and effects sends all work in basically the same way. Some AUX sections on consoles are designed with their own set of effects sends allowing audio from the original input to be routed to and from a number of parts of the console.

Console Outputs

A console may have a number of different outputs. Some consoles are configured so that all outputs may be used simultaneously; others offer alternatives. Consoles with multiple outputs often have a system of switched buses. Buses are conductors or pathways within the console through which audio and data are routed.

In addition to foldback, AUX, and effects outputs, which are outputs of a specific type, it is not unusual to find main, alternate main, direct, submix, monitor, and a variety of other similarly named outputs. These outputs can be high-impedance, unbalanced outputs or low-impedance, balanced outputs.

Main outputs can range from something as simple as one monaural line out, to several stereo line outputs, to a separate output for each individual input. They are normally placed after the master faders so that changing the level of the master faders changes the level of the outputs. Alternate mains usually run parallel to the mains and serve as an extra set of outputs. Their output is also controlled by the master faders.

Direct outs usually bypass the master faders. Some consoles are equipped with separate direct outs for each channel. Any or all of these outs may be used to make a separate mix for a variety of other purposes, or they may be fed directly to a recorder so that a multitrack recording may be made that is not affected by any changes made to the main console.

CONCLUSION

Having the most advanced studio in the world is no guarantee of great production. Great work is regularly done on the simplest of systems. This fact is often lost on the new producer, who is easily distracted by the fascination of the studio and to whom the studio is an end in itself. Although you may be amazed at your own work, don't let the amazement show to your clients. Their only concerns are how the finished product sounds, how much time it took, and how much it cost. By revealing your amazement, you may cause them to wonder whether you really know what you are doing, or whether you just got lucky. The experienced producer integrates all of the technology seamlessly and sees the studio as a means to an end. By knowing your system, knowing where the signal goes, and knowing the equipment and how it works, you become part of the system. This is the producer's magic.

5

Shaping Audio

The shaping of audio through the use of audio-processing circuits is one of the most subjective areas of audio production. It becomes difficult, if not impossible, to quantify the effects of a producer's decision to add, subtract, or otherwise alter any audio from its original form. For example, would a music group's album, cassette, or CD sell more copies if one knob were turned up just a little? Would a radio or television commercial persuade more consumers to buy the product if that same knob were tweaked a bit in the other direction? Probably not, unless the knob happened to be the on/off switch. Still, the cumulative results of minor adjustments make all the difference in the finished production.

As the realm of audio production continues to unfold before the practicing producer, each person must face an incredibly large number of aesthetic decisions that are more about art and craft than about electronics or science. In addition to knowing whether a device can be connected to a specific place in a system, you must also know if using it there will give you the desired effect. Nowhere in a producer's purview are the decisions more abundant and complicated than in the recording, processing, and mixing of audio.

If the answer to the questions raised earlier is yes, the producer must decide how much or how little of each effect to use. Too little will not be heard. Too much will overpower the sound, distracting the listener and detracting from the finished production. Those decisions lead to the next level of choices, which determine whether the chosen effects are used throughout the entire selection or only on a part of it. Depending on the context, the producer may choose to increase or decrease the amount of effect at any given point to further enhance the mood of the moment.

Because this can be a very complex process, it is next to impossible to make good decisions unless you first understand where and how that equipment can be used. The first step in being comfortable with these kinds of decisions is to know as much as possible about the inputs and outputs of each piece of the system and how they interconnect as well as what options for connection the system allows.

In the strictest sense, any processing device that alters the original waveform of a signal is distorting it. However, because *distortion* normally carries a negative connotation, some people prefer to use the terms *waveform modification* or *audio processing*.

AUDIO PROCESSING

The basic concepts of audio processing can easily be simplified. The many different circuits may be categorized as equalization, gain reduction (limiters, compressors, de-essers, expanders and gates), time-domain effects (reverb, echo, delay), and psychoacoustic effects (aural exciters, harmonic enhancers, spectral enhancers).

The best way to absorb the knowledge necessary to use these circuits intelligently and creatively is first to understand what they do generically. Armed with that information, you can make better decisions as to which of the many similar circuits will work best for you. Each circuit, by its design, will provide different results. Compressors from different manufacturers and even different models from the same manufacturer affect the sound differently. In addition, most controls on all but the simplest of each of these circuits can be adjusted so that the effect ranges from extremely noticeable to extremely subtle.

Most producers establish a collection of favorite pieces of processing gear over a period of time. Their preferences are based primarily on their satisfaction with what a piece does and with their familiarity with its menu of possibilities. Attributes such as flexibility, reliability, and value also figure into the choice, but not as heavily as the menu of effects the producer feels comfortable with using.

EQUALIZATION

Analog Equalization

Through EQ, a producer sculpts raw sound into a finished production or into elements that become parts of a finished production. Analog equalizers are circuits that allow modification of the frequency response of a signal. In their simplest form, equalizers may be designed to block a certain band of frequencies from passing through a circuit. An equalizer that blocks the passage of all frequencies below a desired point is usually called a *high-pass*, or *low-cut, filter*. An equalizer designed to block all frequencies above a desired point is called a *low-pass*, or *high-cut, filter* (Figure 5.1). These filter switches are normally found on both mixers and some microphones.

Typical remedial applications of a high-pass filter are to block, cut, or roll off undesirable lower frequencies, such as 60- and 120-Hz hum from poorly grounded circuits or extraneous acoustic or mechanical noises, such as wind or noise from nearby machinery. Low-pass filters are normally used to block high-frequency noises, such as circuit noise and tape hiss.

Both high-pass and low-pass filters are commonly used in electronic news-gathering (ENG). Because most of the audio is voice only and because most voices have little if any energy below 100 Hz or above 12 kHz, the frequencies above and below these points are often rolled off. Eliminating frequencies above and below the voice frequencies usually makes the voice easier to understand.

The filters are either in or out; there are no in-between settings. To get those in-between settings, you need adjustable EQ, which usually comes in two configurations: step-switched or continuously variable. The simplest form of adjustable EQ is shelving, which is the ability of an equalizer to control the amount of a specific range of frequencies in a linear fashion, usually at the bottom top, or both of the audible frequency spectrum. All frequencies that are part of the shelved area are affected equally when adjusted (Figure 5.2).

Step-switched EQ lets you choose specific frequencies at which a preset amount of roll off will begin. Typically, the choices are from 50, 100, or 200 Hz on the low end and 5 or 10 kHz on the top. Step switches may also be used to control the amount of boost or cut at a particular frequency (Figure 5.3). Continuous, or sweep, EQ allows you to fine-tune the amount of a cho-

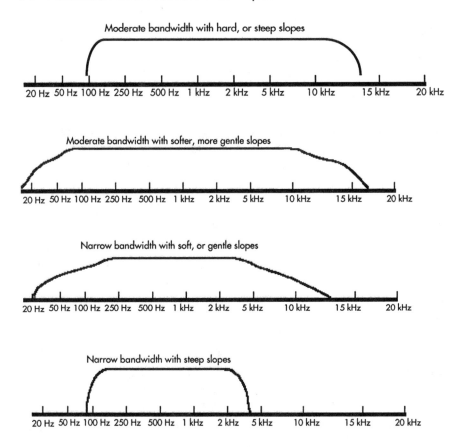

Figure 5.1 Low-cut and high-cut filter effects. Low-cut and high-cut filters may be designed so that the frequency at which they begin to roll off is adjustable. The rate or slope at which they roll off may also be adjustable. The high and low frequencies are also independently adjustable.

sen frequency or range of frequencies with the turn of a knob for even more precise control.

The graphic equalizer provides greater control by dividing the audible frequency spectrum into a number of bands, which are usually measured in octaves or fractions of octaves. The greater the number of bands, the more precise the control. The musical concept of octaves becomes apparent when you look at the face of a graphic octave equalizer. The bands—25, 50, 100, 200, 400, 800, 1600, 3200, 6400, and 12,800 Hz—are each one octave apart. To calculate the octaves of any frequency, simply divide or multiply by two (Figure 5.4).

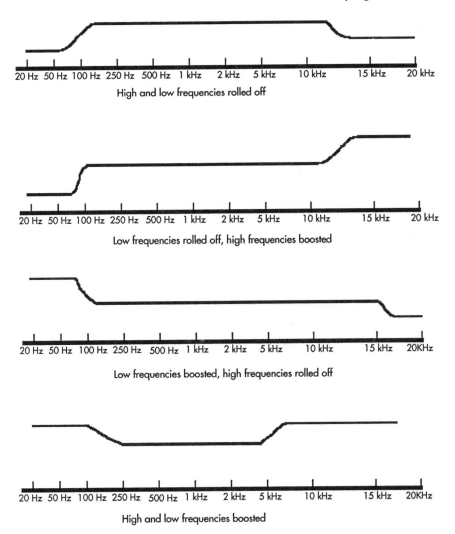

Figure 5.2 High and low shelving equalization effect.

A parametric equalizer provides even greater control of the parameters of the frequency spectrum. It allows you to choose a band of frequencies, the width (or Q) of that band of frequencies, and how much of that band passes through the equalizer. Most parametric equalizers offer control of the entire audio spectrum, using three or four bands that overlap slightly (Figure 5.5).

One unique application of the parametric equalizer is its use in finding problem frequencies. If, for example, you hear a frequency or range of frequencies that seems to ring out louder than the others, start by deciding in

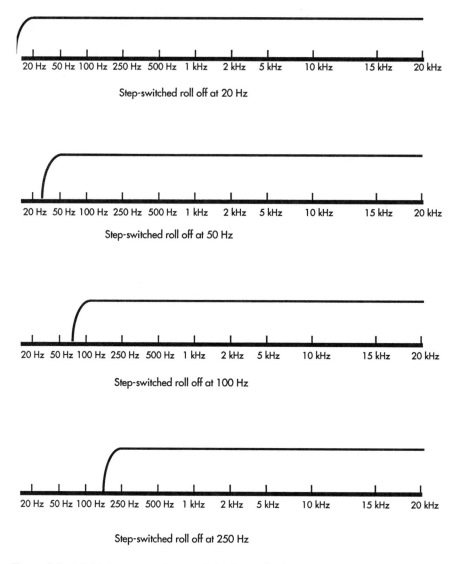

20 Hz 50 Hz 100 Hz 250 Hz 500 Hz 1 kHz 2 kHz 5 kHz 10 kHz 15 kHz 20 kHz

Step-switched roll off at 20 Hz

20 Hz 50 Hz 100 Hz 250 Hz 500 Hz 1 kHz 2 kHz 5 kHz 10 kHz 15 kHz 20 kHz

Step-switched roll off at 50 Hz

20 Hz 50 Hz 100 Hz 250 Hz 500 Hz 1 kHz 2 kHz 5 kHz 10 kHz 15 kHz 20 kHz

Step-switched roll off at 100 Hz

20 Hz 50 Hz 100 Hz 250 Hz 500 Hz 1 kHz 2 kHz 5 kHz 10 kHz 15 kHz 20 kHz

Step-switched roll off at 250 Hz

Figure 5.3 Multiple curves of step-switched equalization.

which band the sound occurs. Turn the frequency boost control of that band all the way up. Then sweep the band of frequencies until the sound you are listening for jumps out even more. Experiment with the bandwidth, or Q, control to determine how wide the bandwidth of the objectionable sound is. Once you have determined its parameters, reduce the gain until those frequencies sound proper in the mix.

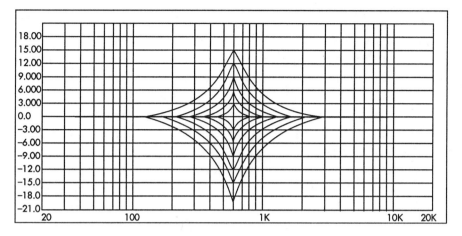

Constant-Q Boost/Cut Performance

Figure 5.4   Singleband graphic equalization.

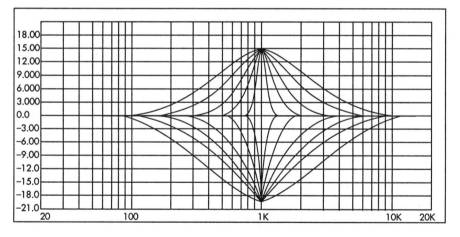

Boost/Cut Performance vs. Bandwidth Settings

Figure 5.5   Singleband parametric equalization.

The operation just described is an example of subtractive EQ. The frequencies that stood out too much were reduced. Additive EQ works the opposite way. In the above example, the producer might have chosen to boost some of the other frequencies to balance the frequency response. Consider that turning down the bass and/or turning up the treble can have similar effects on the overall EQ balance. For example, a passage of audio that is judged to have too many low frequencies can be corrected either by reducing some of the lower frequencies or by increasing the middle or high frequencies. It is usually better to take a little off the bottom and to add a little on the top than it is to simply boost the higher frequencies or cut the lower ones. The idea is to get as consistent and linear an envelope as possible. Boosting any range of frequencies too much will cause those frequencies to overdrive the circuits through which they pass, resulting in distortion.

Another type of distortion that results from even moderate use of EQ is phase distortion. To the experienced ear, phase-distorted sound is smeared and less distinct. Although the better and more expensive analog equalizers are designed to minimize the effect of phase distortion, every EQ device imparts some of this by-product on the audio that passes through it. Over time, most producers learn to identify phase distortion. Their decision as to how much EQ to use is then determined by both the change in frequency response and the amount of phase distortion created by the equalizer.

Digital Equalization

Unlike analog EQ, which changes the waveform of audio by combining out-of-phase and in-phase parts of the frequency spectrum to the original signal, digital EQ achieves its effect by adding or subtracting 1s and 0s at the appropriate place in the digitized audio bitstream. Remember, however, that digitizing audio is not a cure-all for bad audio. If the analog audio is noisy to begin with, its quality will not be improved by converting it to digital.

Typical Equalization Applications

After remedial problems have been taken care of, the producer can then concentrate on applying EQ to enhance the various elements of the mix. Let's consider a production that consists of a voice track and a stereo music track. The same principles apply to a mix that is totally instrumental. Keep the following points in mind when making EQ decisions:

1. Is the music track produced in such a way that all frequencies sound normal and natural? If not, the music track may need to be equalized to achieve a proper tonal balance. Be aware of the various imbalance possibili-

ties. Think in terms of the music existing in three separate bands: lows, mids, and highs. Does the new track have overly pronounced low or high frequencies? A booming bass line or an overabundance of cymbals, synthesizers, or other high-frequency-producing instruments will make the voice more difficult to hear. Conversely, if the low frequencies are unnaturally quiet, the music track will lack power. If the high frequencies are lacking, the music track will sound dull and will lack bite and sparkle. A deficiency of middle frequencies, although not as noticeable as a lack of lows or highs, will also reduce the power of the music track. Too many middle frequencies will make the voice very hard to hear because the human voice is centered in the middle frequencies.

Some producers deliberately reduce the amount of middle frequencies in a music track to make room for the voice track. This is particularly important if the music track contains a lot of brass, woodwind, guitar, and keyboard instruments or any other instrument that plays in the same frequency range as the voice. If you have any doubts about the EQ you have applied to a music track, compare it with an existing music track with which you feel confident.

2. Does the voice track cut crisply against the music track without sounding overly bright, or does it suffer from being overly bassy? If the voice has more bass than the music track, you are probably in trouble, unless you are mixing a male voice with a flute solo or some other instrumentation that is above the natural range of the voice.

One of the most common mistakes a producer can make is to improperly increase the bass frequencies of a voice to give it more power. Because the power in the human voice does not exist in the lower frequencies, the voice becomes "boomy" and actually loses power and articulation. If the voice has too many low frequencies, start by decreasing the 80- to 120-Hz range slightly, or move the microphone away from the performer. By their design, some mikes increase in bass response as they get closer to the source of the sound. This phenomenon is called *proximity effect.* If the voice still does not stand out, try increasing the frequencies between 800 Hz and 2 kHz. Increasing these frequencies won't make the voice sound brighter, but it will add power and presence. Finally, if the voice sounds muffled and reducing the lower frequencies has not achieved the desired effect, try increasing the frequencies in the 3- to 5-kHz range. Increasing these frequencies will definitely make the voice sound brighter, but too much of an increase will make the voice sibilant.

Some voices are naturally sibilant. The sibilants in the English language are the *s, z, sh, ch,* and *j* sounds. A certain amount of sibilance is natural, and the amount varies from voice to voice. It is only when sibilance is overly pronounced that problems occur. In overly sibilant speech, the hissing sounds are noticeably louder. Because they are louder, you may find that the higher energy of the sibilant passages pushes your system into distortion, which makes the sound even more annoying. In some cases, EQ can be used to reduce the high frequencies created by sibilance. If this doesn't work, using a microphone with a more reduced high-frequency response may help. If the voice is already recorded, you may still be able to control it with a de-esser circuit, which is described later in this chapter.

Similar sculpting techniques can also be used to clarify a mix composed of a number of different instruments. For example, the upper notes of a bass guitar passage may be in the same range as the lower notes of the piano arrangement. Unless the composition is specifically arranged to allow both parts to exist simultaneously, the result may be a confusing overabundance of tones that slur or muddy the piece. In this situation, the producer may choose to roll off some of the higher frequencies of the bass guitar while rolling off some of the lower frequencies of the piano part. The producer's concept will regulate how much sculpting is required.

As the number of instruments in the mix increases, more sculpting is required to trim away the parts of each sound that may interfere with other sounds. The sculpted sound may seem very unnatural when "soloed" or heard without the other pieces of the arrangement. Remember, however, that you are making room for each piece to be heard clearly. Do not get caught up in the process of trying to make an instrument sound natural when played by itself.

Just as it may be important for a producer to clarify a mix by using EQ to separate its elements, it may also be important to overlap or combine similar frequency ranges of sounds or instruments. How much to separate, how much to blend, and when to blend are all important decisions that are based on the producer's understanding of the overall concept of a particular production. Although the concept may be somewhat abstract, its effect on a production is very real and very audible. Figure 5.6 shows three of the many possible examples of the concept. Part A shows complete layering, part B shows partial layering with some separation, and part C shows a tangential approach with no layering.

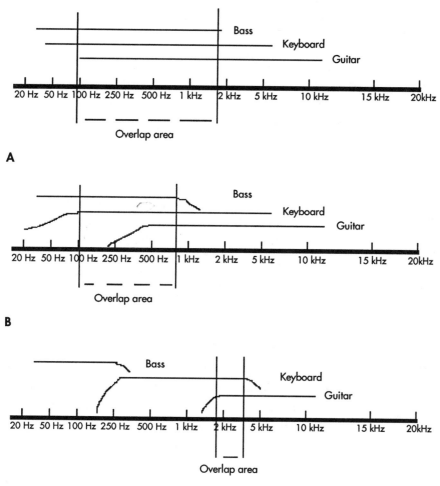

Figure 5.6 Layering of equalization: (A) complete layering (no equalization), (B) partial layering (some equalization), (C) minimal layering (more radical equalization).

Bandwidth Control

Bandwidth control is the use of EQ to determine the highest and lowest frequencies of an audio source. If you are producing music that will be transferred to a compact disc, you will most likely want to have the full 20-Hz to

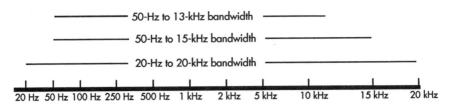

Figure 5.7 Representation of bandwidths from 20 Hz to 20 kHz, 50 Hz to 15 kHz, and 50 Hz to 13 kHz.

20-kHz bandwidth to allow all of the highest and lowest sounds to be heard (Figure 5.7).

As in all mix decisions, the producer must remember end-use considerations: What effect will the playback system have on the EQ adjustments? It is not at all unusual for a producer to purposely limit the bandwidth of parts of a mix or an entire mix to less than the full-frequency response of which the equipment is capable. By reducing the very lowest frequencies and the very highest frequencies, the producer can raise the overall level of the audio that remains. Although the mix will lack full fidelity, it will be louder than a recording that does pass the lowest and highest frequencies. Variations on this technique are commonplace in radio and television audio production.

The bandwidth of both frequency modulation (FM) and television audio is 50 Hz to 15 kHz. Allowing frequencies above or below those frequencies is not only futile, it can actually reduce the overall loudness and clarity of the broadcast signal. Because there is a considerable amount of processing before the stage in most FM and television audio chains during which bandwidth limiting occurs, the frequencies above and below the 50-Hz to 15-kHz bandwidth may have a negative effect on the way the gain reduction devices handle the audio. By sculpting the audio to fit the bandwidth, you make it easier for compressors and limiters in the broadcast audio chain to handle the audio.

DYNAMIC GAIN CONTROL

As its name implies, dynamic gain control is the automatic control of the output of a system. Before these circuits were invented, engineers achieved a similar, but less precise, control by manually raising or lowering the gain at some point in a system. This practice, called *riding gain*, still works when changes are slow enough to be anticipated by an engineer or producer. Even today, some of the best audio producers ride gain meticulously on music and voice tracks in real time to make up for what they hear as dynamic inconsistencies, or very small volume changes, that work against the other elements in the mix. If the voice talent moves slightly away from the microphone or begins to run out of air during the reading, the producer may be able to compensate for the drop in volume level during the mixing process. Increasing the volume of a voice track just a couple of decibels on just one word or phrase can sometimes make a major difference in the impact of the message.

A producer can also ride gain on music tracks. For example, a piece of music may have sections where more instruments are added to the mix or where the existing instruments become louder. In some cases, lowering the music bed slightly successfully prevents these louder music passages from overwhelming the voice track. However, at the point at which the changes are too numerous or too complicated or happen too quickly, it is preferable to use gain reduction circuits, such as limiters and compressors. Because of the flexibility they provide, most producers use gain reduction devices in addition to manual moves to achieve the proper packing of a mix.

Dynamic Range Circuits

All of these devices in some way control the dynamic range of the audio that passes through them. As mentioned earlier, dynamic range is simply the spectrum, or range, of sound from loudest to softest. Audio that falls below the working dynamic range of a system becomes lost in circuit noise. Recordings made below the working dynamic range of an analog tape recorder are also noisy due to tape hiss. Because many analog-to-digital converters use only the full number of bits on the loudest signal that passes through them, even digital audio recordings can suffer lack of definition if recorded too low. Each device that originates, passes, or receives audio has its own working dynamic range. In contrast to compressors, limiters, and de-essers, which reduce dynamic range, downward expanders and gates increase dynamic range by turning down the level of audio that falls below a set threshold.

Gain Reduction Applications

Although compressors and limiters were originally designed to keep audio levels from exceeding the usable dynamic range of a system, another by-product of their function is increased loudness. Compressing and limiting audio are ways of compacting it so that it becomes more dense. The more consistently dense it becomes, the more power it possesses. At some point, however, there is a tradeoff between power and form. Figuratively speaking, you could probably compress enough ballerinas together to make a sumo wrestler, but at some point they would stop looking and behaving like ballerinas. The more you know about how each gain reduction device works, the better you can determine how much of each you should use.

There are several typical applications for compression and limiting in the production studio in addition to creating a denser and louder sound. Music tracks that have too many loud and soft passages can be packed to make them more consistent so that they will neither interfere nor drop out when used as a bed with a voice track. In the same way, voice tracks can be made more powerful. The concepts are not that difficult, and once you learn them, you can apply the knowledge to any piece of gain reduction gear you encounter.

Gain Reduction Basics

As shown in Figure 5.8, if the meter fluctuations of a section of audio show that the quietest parts of the passage register –50 dB and the loudest parts register 0 dB, that section of audio is said to have a dynamic range of 50 dB. If 10 dB of compression or limiting is applied to this 50-dB range, the new dynamic range is reduced by 10 dB and becomes 40 dB. The metering on the console normally reflects this. The signal that fluctuated between –50 dB and 0 dB without processing will now fluctuate between –50 dB and –10 dB after 10 dB of gain reduction is applied (Figure 5.8).

Because the louder passages are reduced in volume so that the audio now peaks at –10 dB, the overall level after gain reduction can now safely be increased by 10 dB so that it peaks at 0 dB again. Now the meter fluctuates between –40 dB and 0 dB. In addition to turning down the louder portion of the audio by 10 dB, we have now increased the overall level so that the quieter portions of the audio are 10 dB louder. Because the average modulation level has been increased, the audio is perceived as louder.

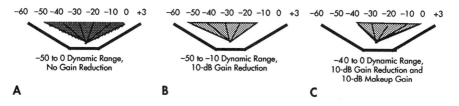

-50 to 0 Dynamic Range, No Gain Reduction

-50 to -10 Dynamic Range, 10-dB Gain Reduction

-40 to 0 Dynamic Range, 10-dB Gain Reduction and 10-dB Makeup Gain

A **B** **C**

Figure 5.8 Changes in meter readings caused by the use of gain reduction devices.

The most commonly found controls on limiters and compressors are attack time, threshold, compression ratio, and release time. As these circuits become more specialized, more controls are added. Some compressors and limiters incorporate special circuits that process transients separately from the main part of the audio. Some have separate trigger outputs that use a signal derived from the audio that passes through it to control another processor. Some split the audio spectrum up into bands, allowing each band to be processed separately before being recombined at the output.

Compromises begin to occur at some point in most gain reduction devices. The tradeoff is usually between average loudness and fidelity. Knowing what effect the controls have on the audio will allow you to achieve the desired effect.

Attack Time

The attack time of a gain reduction device is the time in milliseconds (msec) or microseconds (μsec) required for the circuit to respond to audio that passes through it. Fast attack times may be necessary to prevent the audio from exceeding the headroom of the circuit. The faster the attack time, the more quickly the circuit responds. Transients are those portions of audio that are high in level but last a very short period of time. Percussion instruments, such as drums and cymbals, have very loud transients when they are first struck. By comparison, the human voice has very few, as Figure 5.9 illustrates.

The more quickly the attack time is set, the more quickly gain reduction is applied to the transients. Attack time can be set so slowly that the leading edge, or first part, of a section of audio passes through the circuit before the circuit can be activated. This means that the louder transient will pass through without being processed. As the attack time is set to increasingly faster settings, the circuit will respond more quickly, catching and reducing the volume of any audio that passes through it. At some point, the

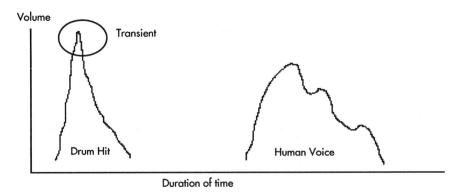

Figure 5.9 Two waveforms: drum hit and the human voice.

reduction in level of transients results in a loss of high frequencies, and the audio sounds muddy. Fast attack times are sometimes used instead of EQ to reduce the high-frequency content of an audio passage.

Some compressors and limiters are designed with preview circuits. The audio that enters the device is delayed by a number of milliseconds. When the preview circuit senses a passage that exceeds the threshold of the circuit, a signal is sent directly to the gain reduction stage. By the time the audio hits the gain reduction stage, the circuit has configured itself to handle the passage with minimal distortion and side effects.

Threshold

The threshold control is used to set the level above which gain reduction occurs. Since audio below the threshold is not affected by the gain reduction circuit, the threshold lets you determine how much of the audio is subjected to gain reduction. When using a compressor followed by a limiter, it is a common practice to work the compressor so that it is constantly compressing the audio that passes through it. The limiter is then set to catch the peaks that are too quick for the compressor. Limiter thresholds can also be set so that they are constantly limiting. When used this way, the limiter becomes a high-ratio compressor.

Compression Ratio

The compression ratio, or slope, is the number that algebraically represents the gain reduction action taken by a compressor or limiter. It is the expression of the difference between increases of signal to the input of a compressor or limiter and those measurable at its output. A compression ratio of 4:1 means that for every increase of 4 dB at the input, the output will only increase 1 dB. It is generally accepted that ratios less than 10:1 are compression ratios, while those greater than 10:1 are limiting ratios. The compression ratio is what differentiates a compressor from a limiter. In simplest terms, a limiter is a faster and more severe compressor. Some gain reduction devices are capable of either compression or limiting, but not both. They have one ratio control that ranges from 1:1 to 20:1. These ratios reflect the amount of gain reduction applied to the parts of the audio that exceed the threshold as it passes through the circuit.

It may be easier for you to think in more visual terms. Figure 5.10A shows a waveform of a single piano note that is not compressed or limited. The leading edge of the waveform, when the note is first struck, is louder. The note then dies out. Part B shows the same note. The dotted line shown in part B represents the threshold of the compressor or limiter. As the threshold is lowered, meaning that more gain reduction is applied to the signal, several things happen. First, the attack of the note is prevented from reaching its peak volume. This makes the shape of the piano note change from a peak to a plateau, as shown in part C. Part D shows the results of make-up gain applied after the compressed note. As before, the quieter parts of the note are louder, and the gain of the louder parts of the note has been reduced. Because the quieter parts of the note have been turned up by the action of the compressor or limiter, the note rings out longer. Producers often use this technique to give a sound more sustain.

Figure 5.11 shows a more graphic representation of the effects of compression, limiting, and expansion. At a 1:1 ratio (points A, B, C), the device has no effect on the gain of the audio that passes through it. As a threshold and a ratio of 2:1 are set (points A, B, D), the output of the compressor increases only 1 dB for every 2 dB it measures at the input. As the angle of the slope increases from 3:1 to 20:1, less of the audio that is past the threshold setting will be passed to the output.

As with most effects circuits, there are tradeoffs. At some point, the amount of gain reduction used will change the sound of the note or of any audio that passes through the circuit. The amount varies from device to device, depending on how each one processes the transient parts of a waveform, and by how the controls are adjusted. Because transients have a lot of high-frequency information and because they are briefly a lot louder than

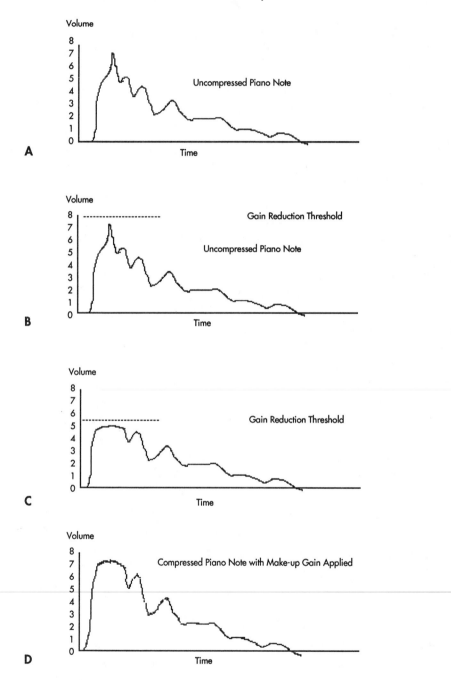

Figure 5.10 Figures (A)–(D) show the effect of applying compression and make-up gain to a piano note.

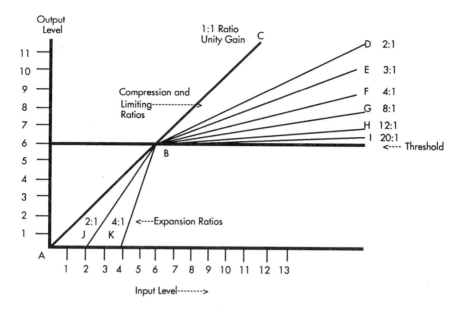

Figure 5.11 A graphical representation of the mathematical functions of compression, limiting, and expansion.

the rest of a sound, attempts to reduce them with compression and limiting normally result in the loss of crispness, punch, and character.

Release Time

Another common control on a limiter or compressor is the release time adjustment. Release time is the amount of time after a signal falls below the threshold before the audio is restored to its unprocessed state. Faster release times usually result in louder audio, especially when used with faster attack times. Again, there is a tradeoff. Past a certain point, the combination of fast attack and release times and high ratios begins to degrade the audio.

De-essers

A de-esser is a very specific type of gain reduction circuit. It is a frequency-selective limiter that is usually designed to limit the frequencies between 3 and 10 kHz. It is at these frequencies that sibilance occurs in the human voice. If changing microphones or simple EQ doesn't reduce the offending frequencies, you may find that a de-esser will. Male sibilance usually occurs in the lower half of the 3- to 10-kHz range, while female sibilance occurs in the upper half. Most de-essers are tunable to the degree that you can adjust them to the area of the offending frequency range. You then apply limiting to that band of frequencies until it is reduced to a more reasonable level.

Adjusting the controls of a de-esser requires a good ear and a delicate touch. With too much de-essing, the track becomes overly limited, causing it to sound dull and muffled. With too little, the sibilance remains.

Expanders and Gates

Unlike compressors and limiters, which are used to control the loudness of a passage above a threshold, expanders and gates are used to control audio after it falls below a threshold. Both expanders and gates are used to reduce any apparent noise in a system. They do this by automatically turning down or off the gain of whatever stage of the system precedes them as soon as the audio falls below the threshold. Expanders usually work more slowly than gates, shutting off the flow of audio that passes through them on a gradual curve. The threshold is usually adjustable on both expanders and gates. The threshold is normally set to allow the quietest section of desired audio to pass above the threshold. Some of these circuits allow you to adjust the rate at which the gain is reduced; others offer only a fixed rate.

Gates are usually designed to close more quickly than expanders. They are often used when more than one microphone must be opened, such as during a music recording session or a live concert. Drum mikes, for example, are often gated to keep other sounds from leaking into the mix. Gates are also used during events like congressional hearings, teleconferences, and church services, where a number of mikes are used to amplify or record the voices of different people at different times.

Without gates, mikes will pick up the specific source to which they are assigned as well as the sound from nearby sources. Because the other mikes are indirectly picking up the sound source, the overall mix will sound hollow and "roomy." The idea is to set the threshold of each gate so that it opens only when its primary source, such as the person assigned to that particular mike, speaks into it. In addition to reducing circuit noise, gates provide an additional benefit to live public-address (PA) applications. Because each mike is on only when it is being used, occurrences of feedback are greatly reduced. This means that the overall level of the system can be increased without causing feedback.

The adjustment of the threshold of an expander or gate requires special attention. If the volume of the primary source varies, the threshold must be set so that it opens on the softest sound. Otherwise, the gate will remain closed, and that portion of the sound will not be heard. When the human voice is the source, it's a good idea to check the threshold when words with soft beginnings—*m, w, n* and *a*—are spoken, especially when they are preceded by a pause. At that point, the gate or expander will be closed and waiting for a sound above threshold. Words beginning with soft sounds may not rise past the threshold quickly enough to open it properly. They are more likely to be clipped than words beginning with *t, p, k, d,* and other hard consonants.

TIME-DOMAIN EFFECTS

There are quite a few reverb and delay processors. Each usually has a number of programs that can be called up to simulate different environments. Armed with a little basic knowledge, you can get the sound you want quickly.

Reverb, echo, delay, flanging, chorusing, and acoustic simulation—whether analog or digital—all fall into the category of time-domain effects. These devices combine delayed portions of the source audio with the original to achieve their effects.

Digital audio processing has made possible an entire generation of time-domain effects. These effects are the result of computerized acoustic analyses of existing spaces,which result in complex algorithms, or mathematical formulas. These formulas attempt to replicate the original reflection patterns that give a space its sound. Once the analog sound is digitized, the bits are applied to the algorithm formulas. If the computer chips that perform the number crunching compromise some of the values by quantizing or rounding off figures, the resulting sound falls short of the original model. Typical by-products of quantization are noise and a metallic ring on the trailing edge of the affected audio.

Another deficiency of low-tech effects boxes is a reduced audio bandwidth. Although the original signal fed to them may have a bandwidth of 20 Hz to 20 kHz, the output, or return of the effects device may only be 50 Hz to 15 kHz, or even less. Because effects return levels are usually low in a mix, this may or may not make a difference in your work. However, it may help to explain the difference in price of seemingly similar effects boxes.

Mono, Stereo, or Both

Another important consideration is whether the device is mono or stereo at both its inputs and its outputs. Although a nice feature, stereo inputs are not as important as stereo outputs. Unlike mono outputs, stereo outputs usually spread the effect across the stereo spectrum, creating a more real sense of space than mono does. In some cases, even though an effects device may have stereo inputs, some of its programs require only a mono input. When this is the case, the audio feeding the left input of the device is usually affected, and the right input audio is ignored.

If a stereo effect is not required and the effects box you are using is capable of independent two-channel operation, you may be able to use the effects box as two mono units, or you may be able to chain two effects together in series for the same signal. With the effects box configured as two mono units, you could run a voice or instrument through a reverb program on the first channel and another voice or instrument through a delay or echo program on the second channel. If your console has two separate auxiliary sends per channel, you could use send 1 to feed the reverb effect and send 2 to feed the delay or echo effect. Although using too much of one effect or too many different effects can produce a very garbled sound, using just the right amount of each can result in a subtle but complex sound that piques the listener's interest.

When using stereo effects, check the final mix in stereo and mono. Many so-called stereo effects programs achieve their effect by simply flipping the phase, or polarity, of the left channel to create the right channel. As long as you're listening in stereo, there's no problem. However, if the stereo is combined to mono, the effect may cancel out, making it inaudible.

Reverb and Echo

Although the terms *reverb* and *echo* are often used interchangeably, they are very different. A reverberated sound rings out; it is perceived as one long sound that eventually dies out. Reverb time, or decay time, is the amount of time required for the reverberated sound to fall below a certain volume, usually −60 dB. The longer the reverb time, the longer the effect rings out.

Echo waves have more discrete and individual reflections. Say "Hello" into an echo device, and you hear the word ring out as individual repeats. Echo effects are usually designed so that the user can control the time between the original sound and the first echo as well as the number of echoes heard and how loud they are. It is the rate at which the reflections occur that tells the ear the size of the space in which the sound is created. Simple echo programs with quick delay times (relatively fast reflections) give the illusion that the speaker or sound is occurring in a small room with very reflective walls, like a bathroom. Longer delay times (slower reflections) give the illusion of a larger space.

Both reverb and echo are complex types of delayed sound. *Delay* simply refers to the difference in time between the original and the reflected sounds. To have a delayed effect, the circuit combines the "dry," or unprocessed, audio with the "wet," or processed audio. Especially with chorusing, flanging, and echo effects, the source audio must be present as a reference or the effect is lost. It is the combination of the original and delayed audio that produces the effect. The slight differences in timing between the source and affected audio create the hollow or tubular sounds normally associated with these devices.

When creating effects with reverb and delay, you must be able to remain focused on the primary sounds in the production. Effects that are too radical draw attention away from the primary sounds, which is especially bad if the primary sound is an announcer's voice. The best way to use echo and reverb in a voice-over production is to use a mix of several different types at very subtle levels rather than just one type. Effects should augment, but not overpower, a production.

There are several ways to achieve this result. If you always record your voice tracks dry, or without effects, using multiple effects requires the use of at least two separate effects boxes or an effects box that contains several programs that can be chained in series. Some devices only delay, some only reverb, and some do both, but not at the same time. Still others allow reverb, delay, panning, and other effects to occur at the same time. In fact, some circuits are so versatile that you can choose the order in which the effects occur. For example, a chorus can feed a reverb that is then fed to a delay. A typical application would be to add just enough reverb first to fill up the combined sound of the voice and music, but not enough to make the voice hard to understand.

Complex audio such as music or busy ambiences takes up space. In doing so, it masks the reverb and other effects applied to some sources, making those effects less apparent. The amount of reverb that sounds just right on a soloed voice track probably will not be enough when mixed with music. After you're happy with the sound of the affected mix, add a small amount of chorusing, flanging, or echo. The amount should be very small so as to add to the thickness of the mix without getting in the way of primary elements, such as the voice track. Although listeners may not be consciously aware of your efforts, they will recognize that there is something very different about what they're hearing, and they will be compelled to listen.

The pacing and rhythm of the elements of the production you're working on should also dictate the echo repeat and reverb sustain settings used. If the reverb sustains, or rings out, too long, the effect will become garbled and will make what is being said more difficult to understand. If there are too many echo repeats or if they come too close together or too far apart for the pacing of the piece, they will be distracting. Begin by using a delay of 20 to 25 msec on either a voice track or a mixed voice-over. Because the delay is so slight, there will be no audible discrete echo, but there will be enough delay to thicken the sound, making it more apparent. Unlike a boost in volume, which simply increases the energy of the piece by raising its level, 20 to 25 msec of delay adds the sound to itself with the added interest of the slight offset between the dry and the wet signals. In a sense, unlike raising the "vertical" volume by increasing the gain of the track, you are increasing the "horizontal" volume by mixing the sound back in with a slight amount of delay. Although the increased presence may not register on the meters, the track will stand out more. Again, the ear and brain are presented with something slightly different from what listeners expect to hear, especially if most of the audio heard before and after is not affected this way.

Stereo Panning

If the final mix will be in stereo, the producer must understand and master the dimension of the stereo spectrum in addition to frequency and time-domain effects. Just as in choosing when to blend and when to separate with EQ, the producer must also judge where on the stereo spectrum each element of a mix should be placed. In complex music productions, the producer uses the stereo spectrum to arrange or place various instruments, usually with the most important ones, such as lead vocals, placed dead center mono.

Kick drum and bass guitar are often panned to the center because they drive the song and because their relatively low frequencies do not compete with the lead vocal. Other instruments are usually panned a bit to each side to allow room for the vocals, unless they are the solo instrument. Mono returns from reverb echo delay and flanging effects may also be panned to any point on the spectrum. Stereo returns may occur as full-stereo images that are combined with the stereo image created by the unaffected audio.

With enough effects sends and returns and with enough different effects boxes, the possible variations seem infinite. Place the dry vocal in the center, and pan the stereo returns slightly outward of the center. If you use more than one effect, you can chain the effects in a progressively outward fashion toward the extremes of the stereo spectrum. This outward cascading of effects gives the vocal added presence across a larger part of the stereo spectrum. Because you have increased its spatial presence, the vocal does not require as much volume to be heard in the mix. Practically speaking, if you listened to a vocal and music track mixed this way and moved your head from one speaker to the other, the level of the vocal would change very little. Without this panning effect, the vocal would be loudest when your head was directly between the speakers and would decrease in volume as you moved your head toward each speaker. Producers often pan all but the primary elements of a mix so that they do not compete with the primary elements for the attention of the listener. Radio commercials produced in stereo lend themselves to this approach.

Even studios with limited stereo capability can turn out complex stereo masters as long as they have one or two stereo record/playback machines for mastering, stereo inputs for music and stereo effects and a number of pan pots for the other mono sources. Remember that panning a number of mono sound effects or voice drops can create a very powerful stereo image.

Panning Considerations

As compelling as the lure of stereo production is, it is good to remember that all the magic you have created may be turned against you after your master tape leaves the studio. Producers who have experienced the pain of hearing their work compromised beyond belief have learned to take a more conservative approach to the use of wide or radical panning. Commercial producers, for example, seldom pan the primary voices more than slightly out of center. For reasons discussed in the next few paragraphs, it is a gamble to put radical pan moves on the most important elements of a mix.

As with most effects mentioned so far, to use the stereo spectrum to the fullest advantage, the producer must consider the end-use system and the compromises that system may place on the stereo mix. The playback system may be technically incapable of recreating the spread of the original stereo field. In addition, engineers or producers at the playback site may automatically convert the stereo signal to mono to make it less of a problem at their end. In fact, some engineers and producers do not even bother to combine the two channels to mono; they simply take whichever channel sounds better to them. The originating producer should try to ascertain whether the end-use system will be mono or stereo and mix accordingly.

The two most frequent reasons for stereo incompatibility are differences in alignment between the record and playback heads of the different machines and inadvertent phase reversals due to improperly connected equipment. As shown in Figure 5.12, differences in azimuth adjustments between the recording head of the machine on which the piece was recorded and that of the playback head on the playback machine are not as obvious when the piece is played in stereo. However, if at any point the playback of the stereo mix is combined to mono, noticeable phase cancellation will occur that cancels or reduces the amount of high-frequency content of the mix, making it sound muddy.

If you know your equipment well enough and are comfortable with making azimuth adjustments to the playback head of your tape recorder, you can adjust the azimuth of your playback head in hopes of matching that of the record head used to put the signal on the tape in the first place. The simplest way to do this is to listen to the mix in mono while you carefully change the azimuth adjustment on the playback head of your tape machine. Pay close attention to the position of the adjustment tool so that you can go back to the original setting. Rotate the tool one-quarter turn to the left and right of center, and stop when the high frequency content is at its maximum.

If, after this procedure, the playback of your tape machine sounds muddy compared to its input, you may not have returned the playback head to its original position. If this happens, you can realign the playback head

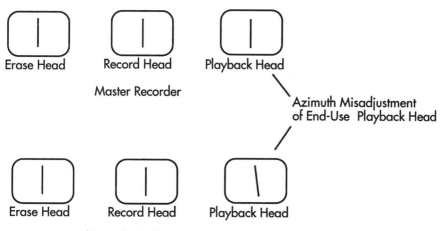

Erase Head   Record Head   Playback Head

Master Recorder

Azimuth Misadjustment
of End-Use Playback Head

Erase Head   Record Head   Playback Head

End-Use Playback Machine

Figure 5.12   Close-ups of two tape recorder head blocks, showing erase, record, and playback heads. The block on the top is the master recorder. The one on the bottom is the end-use playback machine.

relative to the record head by recording material with a lot of high-frequency content and adjusting the playback azimuth for maximum high-frequency content while listening to the tape output as you record. At this point, however, the best thing may be to use an alignment tape to verify the playback head adjustment and then readjust your record head. If you are uncomfortable with these procedures, get an engineer to do it.

Another reason for a stereo mix from another studio to sound bad when monitored in mono in your studio is complete phase reversal of one of the channels. Simply put, the two wires carrying the audio from one channel are reversed. If your equipment is connected to a patch bay, you can correct this error by using a phase-reversal patch cord. In an unbalanced system that has only two conductors, the wire that connects from the tip of each plug should be connected to the sleeve or barrel of the other plug. In a balanced system, the two wires carrying the audio are reversed. Be sure to label this patch cord so that it will not be used by mistake.

Ambience

Ambience is naturally occurring sound or effects combined to give the listener the impression that the source is happening in a certain space. Ambience allows the producer to make it sound as though the actors are somewhere other than in a recording studio. Great care must be taken when recording the voices to ensure that the actors' proximity to the microphone and the naturally occurring reverberant nature of the studio do not predetermine the ambience of the recording. If the actor is supposed to be across the room from the listener and the recording of the actor's voice is made with the actor close to the mike, the voice will lack the reflections the ear normally expects to hear from someone speaking from across the room. Simply turning down the gain to make the actor seem to be farther away doesn't work because, although the voice is quieter, it does not occur in space of the proper size. You must either place the microphone some distance from the actor in a room of acceptable size or use one of the many echo or reverb devices to simulate a room of the proper size.

Outside ambiences are the most difficult to simulate in the studio, especially if the actors are supposed to be some distance from the microphone. The voices must be recorded in a studio with as little room sound as possible. Even minor echo reflections can provide telltale audio cues that tell the listener that the voice is in a room and not in a large open space.

The producer also creates ambience by using sound effects and prerecorded ambiences. Location recordings and effects tracks, available from a variety of sound effects libraries, make the listeners believe that the audio they are hearing was recorded in settings other than the recording studio. More continuous elements of ambience—such as wind, water, and traffic—may sound particularly bland when used "as is" from a sound effects library.

To accomplish a more interesting effect, the producer can preproduce a montage of the needed sounds, using a number of different stock tracks. A conversation taking place between two people walking down the street should include street and traffic sounds as well as the footsteps of the two characters. If the conversation is supposed to be occurring on a busy downtown street, the producer may need to add more people and automobile sounds to an existing track to get the right ambience.

More subtle, but equally important, is the motion of the characters. If they are standing still, the perspective should also be stationary. However, if the listener's position changes to keep up with or move away from the actors, the producer must create the appropriate motion for the listener's perspective. This can be accomplished by making additional location recordings in which the mikes are moved to simulate the changes in perspective that the producer is trying to create.

The more complex the ambience, the more difficult it is to recreate. The trick is to use multitrack recording to layer a lot of different sounds at very low levels, creating a rich tapestry of sound and a more vivid image in the mind of the listener. You may be able to recreate motion by panning sound effects so that the listener appears to move past them. When considering the recreation of moving perspectives that move toward or away from the listener, remember that you must increase the pitch of any noise source you approach and must decrease the pitch of any sound you move away from. This pitch change recreates the Doppler effect we hear when we approach, pass, and move away from a stationary sound source, such as a honking car horn.

If dialogue is important, listen carefully to make sure that your ambience sounds are positioned in the mix so that they do not get in the way of the voice tracks. For example, if a door slam or a horn honk occurs on top of an important spoken line, you may have to remix the ambience bed, either reducing the volume of the sound effect or placing the sound earlier or later between words.

Vocal Eliminators

In addition to correcting someone else's phase errors, phase-reversal patch cords are also useful when trying to remove vocals from existing stereo sources for special projects. Remember that you can only do this with a stereo signal which is combined to mono. The degree of success you will have depends on how the source material was produced. The fewer reverb and delay effects originally used on the vocal and the more mono the pan position of the voice, the greater the effect. Use your phase-reversal patch cord to put one of the stereo channels purposely out of phase with the other. If you are not using a patch bay, you can rewire an existing cable or wire up a new one that does the job. Because there is in-phase mono information on both left and right channels, when one is reversed, it cancels out the equal but opposite mono signal from the other channel. Because bass guitar and kick drum are also normally placed in the center of a mix, you will probably notice that they decrease. As you mix the two out of phase stereo signals to mono. Try increasing the low frequency context of the mix to restore the base response.

PSYCHOACOUSTIC PROCESSORS

Because of the competitive nature of the professional audio production market, producers are always looking for an edge that will make their work audibly different and perhaps better than that of other producers. Although *psychoacoustics* is defined as the subjective effect that a sound has on people who hear it, the word is often used in describing an increasing number of effects devices. Manufacturer claims range from increased clarity, greater intelligibility, wider stereo image, and increased spatial qualities to three-dimensional presence from a stereo source.

Fundamentals and Harmonics

To understand how psychoacoustic processors work, you need a basic understanding of the relationship between fundamentals and harmonics. Each sound comprises a fundamental tone and a series of even and odd harmonics. Although some sounds have more harmonics than others, it is the fundamental of each sound that carries most of the energy.

Figure 5.13 shows a pure sine wave generated by a tone generator with no overtones or harmonics. Figure 5.14 shows three instruments playing a note at the same frequency as the sine wave in Figure 5.13. The difference in their shapes shows the difference in harmonic structure and explains why we hear a difference when different instruments play the same note. The harmonic structure of a sound is thus like an acoustical fingerprint. For example, harmonic content is so specific that even if two acoustic guitars hit the same note at the same volume, the difference in their individual harmonic structures allows you to tell one instrument from the other.

Harmonic content, although a small part of a sound proportionally,

Figure 5.13   A pure sine wave.

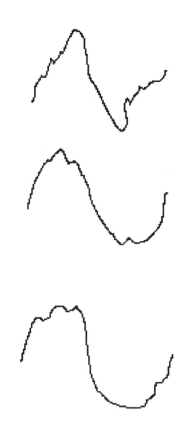

A

B

C

Figure 5.14 Three different instruments playing the same note (A) piano, (B) guitar, (C) flute.

makes a major difference in the overall sound. Because harmonics comprise such a small portion of the total energy of a sound relative to the fundamental, changing them requires relatively little energy. Since harmonics are all higher than the fundamental frequency, adding harmonics normally brightens the sound. A noticeable amount of brightness can be achieved by adding a small amount of harmonics.

Achieving the same effect with an equalizer would be impossible because an equalizer would boost all parts of the audio, including the fundamental. The overall volume of the equalized signal would increase noticeably. Devices such as the Aphex Aural Exciter allow brightness without substantial increases in the volume of the signal.

Aphex Aural Exciter

Not all Aphex devices are equipped with the aural exciter circuit, but the concept for the Aphex Aural Exciter is based on the idea that the circuit analyzes the fundamental and creates new harmonics, which have been filtered or lost. The new harmonics (even, odd, or both, depending on which Aphex circuit you are using) are then mixed back into the original signal.

Suggested uses for the Aphex Aural Exciter include restoring brightness, or high-frequency content, to poorly mixed or recorded material, and increasing the snap, crispness, or attack of elements in a mix, including many sounds from drum machines and synthesizers. Individual elements appear to become louder in the mix when processed by the exciter. Use of the exciter must be carefully controlled. It is very easy to apply too much of the effect, producing very harsh results. One suggestion for achieving the correct amount is to use enough so that the effect can barely be perceived when mixed in but is noticeably missing if taken out after a few minutes of listening.

BBE Sonic Maximizer

The BBE Sonic Maximizer uses a combination of frequency selective delay and companding to achieve brightness and increased definition of the audio that passes through it. The high frequencies appear to have better definition and presence because the middle and low frequencies are time-released slightly after the high frequencies. This delaying technique gives the high frequencies a head start through the circuitry and speakers. In addition to the delayed release of parts of the source audio, the device also has a compressor/expander circuit, which controls the level of the higher frequencies of the processed audio, and a simpler circuit, which allows tailoring of the bass frequencies.

Audio Logic Processors

Audio Logic's PA-88 and PA-86 psychoacoustic processors are both based on the same central circuitry. In simplest terms, each channel is processed with an out-of-phase version of itself that has been slightly delayed. The frequency range at which this effect occurs can be selected by the user. The audible effect is to make high frequencies more apparent.

Bedini Audio Spatial Environment

A close look at the Bedini Audio Spatial Environment (BASE) processor reveals some simple yet sophisticated circuitry. The processor separates the mono information from an incoming stereo source. It then allows the producer to independently vary the gain of the mono signal and to pan it across the stereo outputs.

The phase of the stereo signal is reversed before it is fed to the stereo space control. As you increase the stereo space control, the increasing phase reversal causes the stereo image to widen.

The BASE widening effect disappears when the signal is combined to mono. There is also a slight decrease in what was center channel information when processed stereo is combined to mono. This means that the levels you set while recording in processed stereo, although altered a little, still work if heard in mono.

As with most spatial devices, the effect is more noticeable when heard with headphones. When the stereo space of a stereo music track is increased, it sounds to the listener as though the music is being pushed toward the ends of the stereo spectrum, leaving a larger space in the middle.

When heard over studio monitors, this extra-wide stereo image can be increased to the extent that a hole with no audio forms at the mono point between the speakers. You can fill the hole back up with music by increasing the mono gain control or by mixing in other audio sources—a voice track, for example. You also have the option of panning the mono anywhere on the stereo spectrum.

The amount of separation greatly depends on the nature of the stereo signal you are using. Instruments that are almost on top of each other before processing begin to move apart as the effect is increased. Another noticeable effect on some music tracks is that the intricate percussive sounds seem more defined. The increased definition is partially due to a perceived increase in the 5- to 10-kHz range. Although this increases the overall brightness of the mix, it can also make a normal voice sound sibilant. The widening of the stereo spectrum of an already existing stereo music track presents some interesting possibilities, including being able to move the music out of the center channel area to make room for a voice track without losing overall music volume in a mix.

CONCLUSION

Some producers find experimenting with effects as much fun as producing. Although adding to your bag of tricks and cumulative knowledge is important, be aware that clients rarely appreciate these sonic safaris on their time.

New Technology

6

DIGITAL AUDIO

Without a doubt, the most significant advance in the audio production industry in recent years has been the development of digital audio. Digital audio first appeared in the production studio inside reverb and delay effects, in broadcast studios as time delay circuits, in the recording studio as multitrack recording formats, and finally in a variety of editing and mass storage systems. Digital audio has slowly but surely spread to every part of the audio path between the microphone and the speakers.

Digital audio's primary benefit is its lack of noise. The average signal-to-noise ratio of most analog systems ranges from 65 to nearly 80 dB, a bit more with any of the various noise-reduction circuits. S/N ratios for digital audio circuits typically begin around 88 dB and increase from there. By removing the masking effects of circuit noise and tape hiss, digital technology allows the producer to create recordings of unprecedented clarity.

The distribution of programming via digital satellite broadcast and cable is increasing every day. Some major recording facilities now offer digital-quality satellite services that allow a producer in one city to direct a performer in another city. This two-way satellite link allows the parties to speak to each other over high-quality digital audio channels (Figure 6.1A). Until recently, this sort of phone-patch linking was done over phone lines. Because the phone line audio was not of sufficient quality, the recordings were made on the performer's end and shipped overnight to the producer for final assembly.

If both studios have uplink/downlink capabilities, the producer and talent can now communicate via satellite over high-quality lines, and the recording can be made on the producer's end. A scaled-down version of this scenario requires only one channel of audio. In this configuration, the performance is uplinked to a satellite and downlinked to the producer's studio, where it is recorded. The producer directs the session over a standard phone line (Figure 6.1B).

Some digital systems currently in use transfer digital audio across standard phone lines. The analog audio is converted to digital, and the digital signal is sent over two or three phone lines simultaneously. Decoders on the receiving end recombine the digital data from the multiple phone lines to a single signal (Figure 6.1C).

Basic Digital Concepts

Even if you are not particularly technically oriented, understanding a few basic digital concepts will allow you to make better decisions as a producer. Unlike analog audio, in which we use a microphone to transduce or change the movement of air molecules in terms of volume and pitch into voltages and frequencies, digital audio encodes the audio in a much more complex way. The encoded audio is only a part of a collection of digital information called a *digital word*. Depending on the design of the A/D converter, other digital bits in the digital word are used to make sure that the system performs correctly. The following terms and definitions should help you in understanding the digital domain.

Sample Rate

Put simply, digitizing audio requires that the analog-to-digital (A/D) converter analyzes the incoming analog waveform. It does this by plotting a number of discrete points along the curves of the analog waveform. (This process of analyzing and plotting is generally referred to as *sampling*.) Therefore, as the sample rate or number of times per second increases and as more points are used to define the analog waveform, the accuracy of the converted audio is improved.

Sample rates of 32, 44.1, and 48 kHz have become standard. These figures represent how many times per second the incoming analog waveform

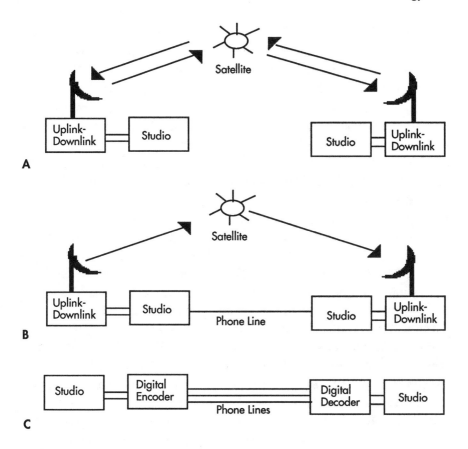

Figure 6.1 (A) Double uplink/downlink. (B) Single uplink with phone line. (C) Multiple phone lines.

is sampled. There is a direct relationship between the sample rate and the frequency response of the digitally converted signal.

A sample rate of 32 kHz produces a digitized signal with a 15-kHz audio bandwidth, which is the same bandwidth as FM or television audio. Even if frequencies above 15 kHz exist in the analog audio that passes through the A/D converter, they will not be sampled, recorded, or reproduced. A sample rate of 44.1 kHz produces a bandwidth of 20 kHz, and a sample rate of 48 kHz produces a bandwidth of 22 kHz.

As you can see, increasing the sample rate results in a wider frequency response and an improved high-frequency response. Simply put, higher sampling frequencies are needed to trace the smaller and more intricate high-frequency waveforms accurately. The more bits you use and the faster you use them, the better the definition and the better the high-frequency response.

The discussion continues regarding how high the sample rate of a signal should be. At the center of the discussion is the issue of harmonics. Even sounds that comprise mostly low frequencies contain harmonics that may be many times higher than the fundamental frequency produced. A bass guitar, for example, that produces a single note of 220 Hz can also create harmonics of that note at 440, 660, 880, 1100, and higher multiples of 220 Hz, perhaps past 15 kHz. Even though the upper harmonics are only a very small percentage of the total energy of the sound, these higher frequencies are a vital part of each sound.

Consider that the highest note on a standard 88-key piano—high C—is 4186 Hz. The first harmonic would be 8372 Hz; the second, 12,558 Hz; the third, 16,744 Hz; and the fourth, 20,930 Hz. In hearing tests using single-frequency tones, only people with excellent hearing are able to hear the 16,744-Hz third harmonic. Although almost everyone except the hearing-impaired can hear the 1100-Hz fourth harmonic of the 220-Hz bass note, less than 1% of the population would be able to hear the 20,930 Hz fourth harmonic of the piano note. As Figure 6.2 indicates, as the frequency of the audio increases, it becomes more challenging to convert it accurately from analog to digital.

Experienced producers and engineers listening to audio over the best systems describe the presence of these harmonics as "hearing more air," or they say that the audio is "more open at the top." Because we know that our high-frequency hearing is responsible for our ability to determine the location of a sound, it is likely that the presence of these higher frequencies is responsible for the perception of air and space. How far do we go to capture upper harmonics? Perfectionists would say that the sky's the limit. Analog circuits capable of passing from 0.25 Hz to 250 kHz exist in today's more expensive consoles. The chain, however, is only as strong as the weakest link. If the audio is passed from the console to a DAT machine operating at a 44.1-kHz sample rate, the highest frequency that can be recorded will be 20 kHz.

Oversampling

Oversampling can be used to increase the accuracy of A/D or D/A conversion. Multiplying or increasing the sample rate requires more sophisticated circuits that run at higher speeds, which reduces the size of the quantizing intervals and improves conversion accuracy.

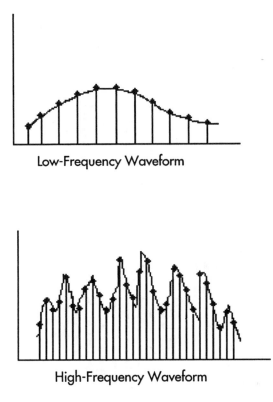

Low-Frequency Waveform

High-Frequency Waveform

Figure 6.2   Two identical sine waves, depicting the concept of conversion accuracy.

Bit Rate

Another element that figures into the quality of the digital conversion process is the number of bits used in each digital word. As Figure 6.3 illustrates, the higher the number of bits used to define a waveform, the more accurate the definition. The current professional bit rate is 16 bits. That is, the portion of each digital word used to define the incoming analog audio is 16 bits long. The higher the number of bits in each word, the better. Even though a system may be rated as a 16-bit system, very quiet passages may only use three or four bits, which results in large inaccuracies in conversion. The true test of a digital recorder is how well it records and reproduces audio at very low levels.

Though 16-bit resolution has become the standard for direct digital conversion, many believe that longer words are needed to maintain fidelity during lower levels and during more intricate processing. Reverb, EQ, echo,

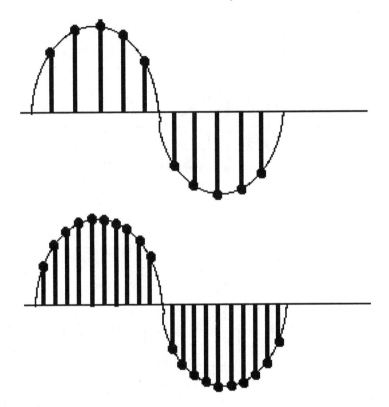

Figure 6.3 Low- and high-frequency waveforms and the relatively higher number of points necessary to accurately plot the highs.

flanging, other time-domain effects, compression, and expansion all involve subtle changes that require more processing power than 16-bit resolution is equipped to handle. Typically, the better digital effects processors convert 16 bit to 20, 24, or 32 bit before processing. In theory, although higher bit count means better definition, the conversion process to and from the different rates must be impeccable, or the sound will suffer.

Quantization

Quantization is the process by which any part of a sampled signal is given a specific digital numerical value. Because the analog waveform is continuous and the digital sampling consists of discrete points of specific values, as the signal is being converted to digital, decisions must be made when analog values fall between two possible digital destinations. The higher the sample rate, the more destinations per space and the more accurate the conversion. As the sample rate decreases, so do the number of destinations. At some point, this rounding off effect produces quantization errors, which reduce conversion accuracy and compromise the fidelity of the audio.

If you find this concept evasive, imagine a scale for measuring weight that reads out digitally in pounds. Even though the actual weight of an object may be 10.9 pounds, the scale will not show that the weight is actually closer to 11 pounds than to 10. If the object being weighed is 1000 pounds or more, 1 pound either way creates an error of only 1/1000. However, if the total weight is only 10 pounds, the error rate is almost 10%.

Reconstruction

Reconstruction is the process of converting digital audio back to analog. Although similar to some degree to A/D conversion, D/A conversion involves "connecting the dots" and reshaping the digital information so that it becomes a fluid analog waveform. As with the A/D process, if a high degree of accuracy is not maintained, the results will be distorted or noisy. Another by-product of improper conversion is the loss of detail during quiet passages, such as when a reverberated sound rings out at the end of a passage. Poor conversion results in premature shortening of the ring out. Poor conversion can also make the audio sound harsh, grainy, or brittle compared to the original.

Some years ago when solid-state analog audio circuits first hit the market, they were criticized by some producers who believed that they sounded harsh in comparison to vacuum-tube circuits, which were then the standard. As research continued, however, the technology advanced to the point where solid-state circuits became capable of amplification without sounding harsh. The development of A/D and D/A conversion is now past its infancy. We can continue to expect improved conversion methods to become commonplace. Do not be led to believe that all digital audio is better than analog. Use your ears.

DIGITAL RECORDING FORMATS

The number and variety of the many storage formats for digital audio are increasing all the time. Current formats include the following.

Tape

Within the linear stationary head tape configuration there are four formats: digital audio stationary head (DASH I and DASH II); Mitsubishi X-80; stationary-head DAT, or SDAT; and the consumer-oriented digital compact cassette (DCC). Within the rotary head configuration, exist the videotape recorder (VTR), C-format, D-1, D-2, D-3, DCT, videocassette recorder (VCR), rotary-head DAT, 8mm digital cassette, Betamax, Super VHS.

Disk Drives

There are three types of disk drives: hard, optical, and floppy. Hard disk drives were originally made by coating aluminum alloy disks with a magnetic oxide medium. Higher-density coatings have been developed that allow more information to be stored. These coatings are electroplated onto the disks. Storage memory varies, depending on the number of disks in a stack.

Optical disk drives include CDs, laser disks, write-once-read-many (WORM) disks, and re-recordable optical disks that are etched and read by laser. Smaller systems for individual CD production are now available.

Although several devices currently use floppy disk drives, the fact that their storage is currently limited to just under 2 megabytes (MB) makes them a limited medium, until data compression technology advances a bit further. The Bernoulli Box is the "big brother" version of the floppy or diskette. With its most recent high-density upgrade, it has a 150-MB capacity.

Random Access Memory

Random-access memory (RAM) uses computer chips to store digital data. Although some chips can be permanently programmed, most RAM memory is considered volatile. *Volatility*, in this case, means that the memory is lost if the power to the RAM chips is turned off or interrupted. Significant advances have been made in RAM technology. The new 4-MB chips have quadrupled the storage space of earlier 1-MB chips.

Interconnect (Electronic) Formats

In addition to the storage formats, there are also a number of different digital audio interconnect formats, including AES/EBU, PCM-F1, PCM-1610 (used for CD mastering), SPDIF/DAT, SDIF, ProDigi/Mitsubishi, EIAJ, JVC, and fiber optic.

Obviously, the greater the number of mechanical and electrical formats, the greater the number of problems you will run into while trying to connect various pieces of digital gear. Even different sampling rates can create problems. For example, if you used a 32-kHz sample rate while recording, you cannot play it back on another machine unless it can play back a 32-kHz sample rate signal. Although some machines will play back sample rates that are different from that which they are set to reproduce, the pitch and the speed of the sound will be altered.

In addition, inputs for any sample rate can be digital, analog, or both. For example, some DAT machines record at sample rates of 32, 44.1, and 48-kHZ through both analog and digital ports. Others may record at those same sample rates through analog inputs but will only allow digital recording at 32 and 48-kHz. This configuration was developed as a way to control the direct digital copying of CDs, which use the 44.1-kHz sample rate.

Another possible impediment to digital piracy is the serial copy management system (SCMS), which was developed to prevent unauthorized recordings from being made. DAT recorders with this system, particularly consumer models, may restrict the number of copies of an original that can be made.

Subcodes

In addition to the digitized audio, digital recorders also record other information, such as Society of Motion Picture and Television Engineers (SMPTE) time code, DAT time code, absolute time, start-ID numbers, program numbers, and an abundance of other information that each machine needs to work properly. Figure 6.4 shows the intricacy of detail needed to provide the large amount of data to ensure a robust, nonfailing system.

Time codes are a stream of numbers—usually measured in hours, minutes, seconds, and frames—that are recorded along with the audio. Usually visible on a machine's front panel, they tell the operator the exact location of the tape on the playback or record head. Time-code-based recording and editing systems use these very precise addresses for automated punching in or out of record or for very accurate editing.

Absolute time, found on some DAT machines, is similar to SMPTE time code but is less precise. Absolute time normally reads in hours, minutes, and seconds and is recorded from the beginning to the end of a digital tape.

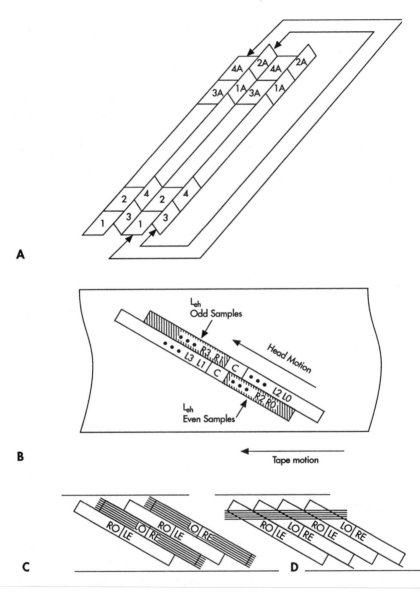

Figure 6.4 (A) In the composite digital format, data from each audio channel are recorded twice so that the copy is at the opposite edge of the tape and read by a different head. This gives immunity to head clogs and linear tape scratches. (B) Interleave of odd and even samples and left and right channels to permit concealment in case of gross errors. (C) Clogged head loses every other track. Half of the samples of each channel are still available, and interpolation is possible. (D) A linear tape scratch destroys odd samples in both channels. Interpolation is again possible. Reprinted with permission of John Watkinson from *The Art of Digital Audio* (Boston: Focal Press, 1989).

Start IDs and program numbers are used on many DAT machines to identify specific selections. When a DAT machine is put into the record mode and a recording is made, an inaudible cue tone, called a *start ID*, is placed on the tape. At the same time, a program number is assigned to that start ID. In most DAT machines, the position of the start ID may be moved after it has been recorded for tighter cueing. Program numbers are sequential. That is, the first program number on a DAT cassette is 1, then 2, and so on. Most DAT machines can be programmed to renumber a series of cuts on a tape if one cut is edited out of the sequence. Many DAT machines can be programmed to play back a list of out-of-sequence cuts. For example, you can program the machine to play cuts 3, 7, 5, 2, and 4, in that order.

Because standards have yet to be agreed upon, the assignment of subcode information varies widely. For example, if you record on DAT machine A, which also records absolute time, you may not see the absolute time on machine B because it is not designed to read absolute time or because the data for absolute time are in a different part of the subcode, which machine B cannot read.

Although you may believe that these complications are more the concern of an engineer than a producer, remember that you cannot depend on an engineer to read your mind. Having a working knowledge of these interfaces is the best way to keep them from coming back to haunt you. All it takes is one unasked question or one erroneous assumption, and your project can be stopped dead in its tracks.

Digital Advantages

Low Noise

One of the many practical advantages that digital audio has over analog audio is its lack of noise. Tape hiss simply is not a factor. Although this increased usable dynamic range allows the recording of quieter sounds, several practical considerations limit how quiet parts of a mix can be.

First, the ambient noise level at the end-use site must be low enough so that the passages that take advantage of the greater dynamic range can be heard. CD recordings of some classical music already exceed the level of practical dynamic range. If the listener is in an automobile on the open road, for example, adjusting the volume to a comfortable level may prove difficult.

If the listener adjusts the volume to hear the quieter passages, the louder passages will be too loud. If the listener adjusts the level for the louder passages, the quieter passages will be lost in road noise. As a result, broadcast engineers often compress the audio so that it stays within a more usable dynamic range (Figure 6.5).

Second, the end-use system itself and the systems through which the audio passes before it reaches the end-use system must be considered. If a commercial is recorded and mixed in the digital domain, it must remain there until it reaches the listener in order to preserve the low noise benefit that digital provides. If, at any point along the way, the digitized audio is converted to analog, noise can become a factor.

Reduced Dropout

Dropout is the complete or partial loss of audio due to inconsistencies in the surface of the recording medium. In the analog domain, that means the tape itself. Dropout can be caused by a bad batch of tape, a tape editor with greasy fingers, or tape that has been stored improperly. Dropout also occurs with very old recording tape and with cheap recording tape. In the worst

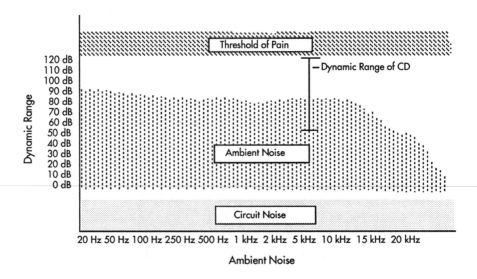

Figure 6.5 Ambient noise versus dynamic range.

form of tape deterioration—shedding—the oxide coating of the tape, which holds the recorded information, breaks down and sloughs off. As the tape disintegrates, so does the recording.

Digital recording systems are designed to reduce dropout. The signal for each moment of audio is recorded in a number of different places on the tape. During playback, if the data at the first address cannot be read due to dropout, the system automatically reads the other locations. Should all addresses be absent or corrupted, error-correction circuits can be used to fill in the blanks based on probability. As long as these lost moments are extremely brief, the error-correction circuits can interpolate or invent a logical stream of data to fill the gaps.

Today, as pleased as most producers seem to be with the sound of digital tape, many still archive an analog copy of their most important works in addition to a digital copy. Their concern is that long-term stability of digital tape has not yet been proven. Until recordings made today have aged a number of years without deterioration, most producers will continue to play it safe.

Transport Stability

Unlike an analog tape recorder, which must pull the tape across the heads with unfailing accuracy and constancy of speed to prevent wow and flutter, a digital recorder with adequate time-base correction can make sense of minor speed discrepancies caused by mechanical problems in the tape transport. Time-base correctors read the relatively simple pulses generated by the digitized audio, move them, and even reshape them, based on what it knows the circuits want to see.

Special attention should be given to digital recordings made with DAT or any other digital format that uses video-style tape transports that automatically thread the tape cassette as it is loaded into the recorder. The automatic loading process sometimes fails to position the tape properly. The recorder's meters show the signal and the transport moves the tape, but the signal never gets to the tape. It does not happen often, but one incident can ruin a whole day's work—and your reputation. An increasing number of DAT machines have extra heads that provide off-tape monitoring, which is also called *confidence recording*. Immediately after the record head lays the signal to tape, a playback head routed to a speaker or to headphones confirms that the audio actually got to the tape.

DIGITAL SIGNAL PROCESSING

Digital signal processing (DSP) is a broad category that includes everything from the action of A/D converters, to the manipulation of audio in the digital domain, to D/A reconstruction. Although some technical impediments to digitally perfect audio remain, most would agree that there are immediate advantages to dealing with audio as a series of *1*s and *0*s rather than as an analog waveform. Compare the processing action of analog and digital equalizers, for example. The analog equalizer is somewhat limited because, in addition to the signal being passed through it, it also passes tape hiss, circuit noise, and phase changes of the audio itself. If correctly recorded, a digital signal has very little circuit noise, has no tape hiss, and does not change the phase of the audio. As a result, EQ changes can be made with a digital equalizer that would create too much noise if made with an analog equalizer.

Inventory and Storage

For the recording studio that does most of its business making radio commercials, audio for video, and other than full-blown 24-track music recordings, the use of a digital audio workstation (DAW) and a DAT, magneto-optical, or other archiving master can greatly reduce tape usage and the amount of storage space needed for tapes. Although it is true that even digitized audio requires disk space, disks can be erased and reused, greatly reducing the need for tape. If you have been using ¼-, 1-, or 2-inch recording tape, the cost and storage space will add up after a while. In comparison, the shelf space required by one DAT cassette, which holds 2 hours of digital stereo, equals the space required by four 10.5-inch reels of 1.5-mil, 2500-foot, ¼-inch reel-to-reel tape stock if recorded at 15 inches per second (ips).

Because digital recorders are capable of recording data as well as audio, multitrack digital projects from digital audio workstations can be archived on a machine as simple as a two-track DAT. In this configuration, the archiving recorder stores the audio information as well as other addressing data. When the contents of the archiving recorder are loaded back into the DAW, the addressing data are used to place each section of audio in the right place on the original track. In more sophisticated systems, recording console settings can also be recalled, eliminating the arduous task of having to recall the position of each knob at the point at which the session was archived.

Nondestructive Editing

One of the main areas in which digital technology clearly outperforms analog is editing. Even though in most situations digital tape cannot be splice-edited the way analog tape can, nondestructive digital editing has several distinct advantages over analog tape splicing.

Because digital audio can be processed so easily by computer, simple editing functions can be done much more precisely than in the analog domain. Accuracy now ranges from 1/128 second to editing of the actual sampled waveform. The beauty of digital editing with a computer is that no data are destroyed. Unlike analog editing, which requires that the tape be cut or that a number of time-code-locked analog reel-to-reel machines be synchronized with an editor, digital editing is achieved by assembling in-points and out-points of particular sections of audio by address and stringing them together. Because the computer that drives the editing system moves at very high speeds, there are no gaps or pauses between sections. Because the in-points and out-points are merely addresses, they can be changed at any time during the editing process without having any effect on the audio itself.

Working with this sort of safety net allows the editor to be much more adventuresome. If an edit does not sound right on the first attempt, the points can be moved very slightly until the edit does work or until the editor decides that the edit will not sound good no matter how carefully it is done.

Electronic Splice Angles

Another advantage of computer-driven digital editing is the ability to create splice angles anywhere from a butt splice to a crossfade of several seconds. Although butt splices and 45-degree splices are commonplace in analog tape editing, as of this writing no commercially available splice blocks have more than a 60-degree splice. Even at a tape speed of 7.5 ips, a 60-degree splice lasts about 1/10 second. Longer splices, which at some point are considered to be crossfades, can make difficult music transitions less noticeable. They also save track space because the second part of a true analog crossfade would have to be laid on a separate track. Because of the power and flexibility of digital audio, a splice of several seconds can be performed on one track, while a splice of 1/100 second occurs on an adjacent track. Figure 6.6 shows the relationship between splice angle, length of splice, and crossfade time.

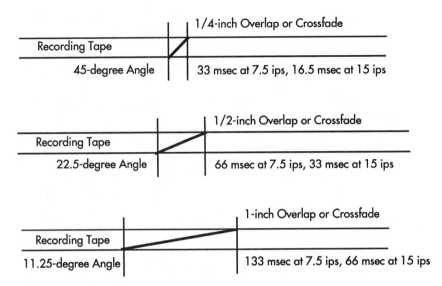

Figure 6.6 Graphic display of different splice angles showing the amount of overlap that each splice produces.

Editing as Craft and as Art

Digital editing becomes even more powerful when performed on a multi-track digital audio workstation with video monitors that show the layout of audio on a number of different tracks (Figure 6.7). Because the tracks are directly in front of you, being able to "see" audio becomes literal rather than figurative. Because these systems allow you to move sections of audio at random, an entire universe of new possibilities opens up. Moving parts of a voice track to achieve better timing with a music track is done very easily. Removing breaths and mistakes and reducing spaces between words are easy. Some editing systems are so accurate that the middle portion of a breath between words can be removed, leaving a very natural sounding short breath. In some of the better digital editing systems, mouth clicks or tongue noises that occur in the middle of a word can be removed without noticeably disturbing the flow of the word. Looping parts of a music track, sound effect, or any other sound bite to fill up a space or correct a timing problem is commonplace.

Editing is a craft. Although many fine points must be understood before you can become a good editor, editing itself does not become producing

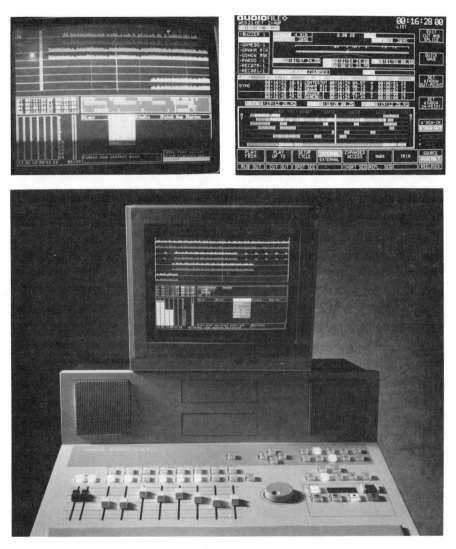

Figure 6.7 A DAW edit screen with audio at different places on different tracks.

until an effective creative overview is applied. Just because you're an excellent bricklayer doesn't necessarily mean that you know the best place to build the wall or to lay the path.

The creative process that raises the craft of editing to that of producing can be the result of divine inspiration or, at the very least, being able to have a visual or sonic image or concept in your mind that can be produced as a recording. For most producers, this complex process involves establishing an overall concept that guides the production. Even so, the creative process often continues throughout the process. At each stop along the way, a producer may find an opportunity to improve on the original concept.

Digital Audio Workstations

There are no fewer than 50 systems on the market today that can be called *digital audio workstations*. They range in price from $4000 to more than $250,000, depending on the options selected. All digital workstations have these things in common: They convert analog audio to digital; they store the digital audio; they use a Macintosh, DOS-based, or proprietary computerized system to display and edit the audio. The more expensive the system, the faster it operates, the more it stores, and the finer its degree of control. Each system operates differently, although some are more similar than others. Many of the early designs required that you first know how to use a Macintosh or a DOS-based computer. Some of the newer workstations have been designed to minimize the computer knowledge required. These newer designs minimize the presence of the internal computers and offer dedicated control panels that resemble the operating controls found on most analog equipment. Continued refinement in this direction will most likely lead to a time when computer operating knowledge, per se, will not be a prerequisite to the operation of a digital audio workstation.

Samplers

Digital samplers usually allow audio to be recorded at a number of different sample rates, from 11 to 48 kHz. The lower sample rates are used to conserve storage space when recording sounds that have little, if any, high-frequency content. The sampler's amount of record time and its width of frequency response are inversely proportional. That is, you get longer record times if you narrow the bandwidth. Some samplers allow you to choose between mono or stereo operation. Some samplers allow you to store just one sound. Others allow you to store a number of different sounds each of which can be played or triggered independently.

Digital samplers are similar in many ways to DAWs. In their simplest use, you can load a voice track, sound effect, or music passage into a sampler and trigger it to play back in much the same way that an analog cart machine does. However, samplers offer significant advantages over the analog cart machines that are typically used for this sort of production.

Once a sound has been sampled, its start and stop points can be edited so that any unwanted parts of the sound at the beginning or end of the sample will not be heard when it is triggered or played. Some samplers allow the recorded sample to be sped up or slowed down without changing the pitch of the sample. This is convenient when you are trying to fit a certain piece of audio into a hole in a production. In fact, if the recording time of the sampler is long enough, you can record an entire voice track and compress or expand its length to fit the required time and space.

Some samplers can also change the pitch of the recorded audio. Vocal harmonies can be created by sampling the primary voice and changing its pitch to the desired harmony interval. Most samplers also allow you to loop the audio so that it repeats the same section continuously. In addition to the possibilities of having a section of music or speech repeat itself at normal speed, some samplers also allow you to change the attack and release times, the pitch, and the speed. Samplers can also reverse the direction of a sample and play it backward. Words or phrases and sounds take on whole new identities. They become textures that can give your production a different sound.

When creating new sounds with a sampler, it is a good idea to record or save a master copy of the finished sound and write down what you did to get it. If the sample is lost, you either have a copy on file or know how to reconstruct it. If you do a lot of work with samplers or other devices that have lots of different adjustable parameters, it is a good idea to keep track of the settings. A small loose-leaf notebook works well. Get the kind that lets you add pages so that you don't run out of space.

Because samplers are normally used to record and play back different musical sounds, many of them are designed to modify the sounds that have been recorded. Some of these modification filter programs are quite extensive (that is, confusing). In addition, many producers have been stymied by the number of buttons and variables found on and in samplers. If you are not already familiar with the waveform modification parameters common in most synthesizers and that particular sampler, you'll have to learn them first. It's a double learning curve that takes a lot of time to conquer.

Operating manuals for samplers and many other circuits are often very poorly written. Although all samplers do basically the same things, the way they do those things is often very different. Therefore, do not assume that you can transfer the knowledge you have acquired from one sampler to another. If you have the time and patience, what you learn will become an envied item in your bag of tricks. If you do not have the time, it is a good idea to find someone to help you get up to speed.

Digital Cart Machines

Digital cart machines provide short- to medium-length record and playback of digital audio, using DAT tape, computer diskettes, and rewritable optical disks as recording media. Because of the limited storage capacity of the disks used in digital cart machines, many of these devices use data compression techniques to increase storage time. Digital audio that has not been compressed requires about 10 MB of storage space for each stereo minute. The formulas created to reduce the data are called *data compression algorithms*. Data compression should not be confused with analog audio compression, which was covered in earlier chapters. Data compression, normally seen as ratios, reflects the increased amount of storage made possible by removing parts of the digital bitstream. Therefore, a 4:1 data compression ratio means that you can store 4 minutes of stereo in the same space that would hold only 1 minute of uncompressed digital stereo.

Several data compression algorithms are currently being used. There is an ongoing dialogue concerning the accuracy of these algorithms and what negative effects they may have on the sound itself. Some listeners claim that they can hear a difference between compressed and uncompressed digital audio. For them, any change from the original is unacceptable. Because the technology is still advancing, it is likely that improved algorithms and higher-density storage mediums will be part of digital in the near future.

MUSICAL INSTRUMENT DIGITAL INTERFACE (MIDI)

Samplers can be operated by a variety of front panel and remote controls or from a keyboard connected to the sampler via musical instrument digital interface (MIDI) cable. MIDI is a machine language that allows different instruments, synthesizers, and an ever-increasing list of audio devices to communicate with each other in much the same way that two computers communicate during data transfers. MIDI information can be recorded, played back, and edited by a MIDI sequencer or by a personal computer with a MIDI interface and the appropriate software.

Simple MIDI, which was originally developed in the early eighties, has already been refined to MIDI time code and "smart" MIDI. In addition to being able to control an increasing number of effects devices, these new MIDI systems are available for the control of console automation. Some consoles are specially built with MIDI-control capability,which allows automated control of level, EQ, pans, mutes, send and return levels, and a variety of effects parameter changes. In some cases, a computer running MIDI-system software is needed. The software is sometimes built into the console. In other cases, existing consoles that were not originally designed to run MIDI can be modified to allow MIDI automation.

In its simplest form, a MIDI sequencer or recorder is an electronic music data box. MIDI does not record audio. When attached to a sound source, such as a synthesizer, it records the performance data of whatever is played on the keyboard. That is, it records when each key on the keyboard is depressed, how long and how hard each key is held down, when it is released, what sound the keyboard is processed to make at that time, and a number of other bits of performance information. This continuous data stream is far simpler to record than the audio of the actual musical notes and takes up considerably less storage space. Upon playback, like a music box, the MIDI data stream replays the performance data, which "plays" the instruments.

Running MIDI sequencing software requires a computer and a MIDI/computer interface. For the additional investment of the computer and interface box, you can see the MIDI information on the computer screen and edit it in much the same way you would on a DAW.

Because it is a data stream, MIDI offers several other distinct advantages. It can be edited and manipulated more easily than audio. If your MIDI arrangement consists of drums, bass, flute, piano, and trumpet, for example, the data can be edited so that the flute performance can be applied to the trumpet sound, and vice versa. Harmonies to a melody can be created by simply making a copy of the melody performance and raising it or lowering it the desired number of intervals.

Tempos can be increased or decreased. If your MIDI composition is 60 seconds long but must be 55 seconds, the entire composition can be sped up without changing the pitch of the music.

If your performance is sloppy, you can quantize, or adjust the timing of individual MIDI tracks or an entire composition to ensure that the notes hit more accurately against the beat. Because quantization is an arithmetic and not a musical process, it sometimes makes things worse. Unless you can undo the effects of your quantization, it is a good idea to make a copy of the track first and then quantize one track at a time. Start with the smallest quantization changes, usually 1/128 or 1/64. If you do not like the results, erase the quantized version, make another copy of the original, and try another quantization value.

Quantization can also be used as a form of artificial intelligence. If you have got a lot of time for experimentation, try quantizing a previously quantized track with a second value. Sometimes you get lucky, and the quantization results in an amazing performance you could never imagine, much less be able to play.

MIDI can also be used to automate console mixing functions. Some console automation systems allow several different mixed versions of the same piece to be stored and interchanged, allowing different sections of each mix to be edited together. If your first mix did not fall apart until the horns came in and your second mix was not really right until after the horns came in, you can edit the two together, choosing the best of each section. The automation will track your first mix until the horns come in, then switch to the other mix automatically.

As MIDI software developers continue to create new applications, they also hone both sides of a two-edged sword. On one side you get the remarkable processing power that a computer can offer, and on the other side you often have to deal with climbing a steep learning curve before you can harness the full power of the system.

CONCLUSION

As the horizon of audio continues to expand, digital audio, the computer, and MIDI implementation can be expected to remain areas of intense activity. The more you know about them and the harder you try to keep up with the amazing amount of change they will bring, the more legendary you will become as a producer.

Until broadcast facility audio chains, network transmission lines, and duplication processes are completely digital and until transmission and reception methods are also digital, the full benefits of digital's lower noise levels will not be realized. Add to that the reality that most commercial producers are trying to make their commercials sound as loud as possible, and the benefit of low noise becomes even less important. These practical ambient and end-use realities combined with improved low-noise analog circuits and a variety of analog noise-reduction systems continue to keep analog audio competitive with digital. It all comes down to the specifications of the system and how good the listener thinks the sound really is.

We are still some years away from the time when analog audio may be completely discarded. There is no reason to replace it as long as it can provide adequate usable dynamic range and is inexpensive, relatively quiet, and easy to use. Also, a large number of producers prefer analog's warmer sound to that of digital. Resistance to digital audio will continue until digital systems can duplicate or improve on those producer's preferences for analog aesthetics.

Symbiosis and Synergy in the Studio

7

Some think of it as bringing a production to life; others call it creating magic. It is the process through which a producer works that results in the finished piece being greater than the sum of its parts — so much greater, in fact, that the sound itself stimulates the listener to form mental images.

Theatre of the mind is a phrase commonly used to explain this phenomenon. It is, however, much easier to identify and talk about than to do. The truth is that theatre of the mind is the result of a complicated and convoluted number of decisions and choices that a producer makes before and during a production session.

It is knowing what to leave out as much as it is knowing what to leave in. In many cases, each decision made along the way establishes the next set of possibilities. The subjective nature of each decision varies from person to person. Even the rapport between people working on a production affects its outcome.

It is difficult to understand the principles of great production because they are dynamic, rather than static. The few hard-and-fast rules that do exist have more to do with operating within the finite limits of the equipment than with the process itself. For example, if you run too much audio through a circuit, the audio will be distorted. Of course, even this rule gives way if a distorted sound is required at the moment.

Even the best producers must please their clients if they are to remain in business. This requires being able to understand what the client wants and providing it in a way that the client is willing to accept. Even if you come up with what you consider the most amazing production ever, unless it meets the client's objectives, its magic exists only in your mind.

Some client requests may be taken very literally, but other clients require the producer to help pin down an idea or feeling that evades simple communication. A successful audio producer, then, must be practiced in the art, craft, and science of production as well as in communications and human relations. A good way to understand how these complex processes work is by listening to experienced professionals talk about how they make magic occur in the studio every day. This final chapter does exactly that, giving you a fly-on-the-wall perspective of conversations with two of the top producers in audio today.

Nelson Funk and Louis Mills have been producing audio for a combined total of nearly 60 years. Because getting them both in the same room at the same time was logistically impossible due to their busy schedules, I spoke to them individually, asking the same questions, and combined their answers.

Nelson Funk has been in the audio production business for more than 30 years. Many of those years were spent running Rodel Audio in Washington, D.C. He began by providing audio for industrial films at a time when studios were still cutting commercials on acetate disks rather than recording tape. Rodel was the first studio in D.C. to offer multitrack recording, using a three-track half-inch machine. At the same time, he was doing an increasing amount of film work, primarily television shows. As New York and Los Angeles began to attract more of the film business, Rodel's audio work continued to increase and included much of the audio work for other local film companies. Nelson Funk soon found it more profitable to shift the emphasis to handling the increasing amount of audio work rather than the business of making films. In the forties and fifties, when radio became a successful advertising medium, advertising agencies began coming to Rodel for production that was better than that recorded at radio stations.

Louis Mills has been in the business almost as long as Nelson Funk. These days, he does most of his session work on an AMS AudioFile digital workstation. Even before he graduated from Johns Hopkins University in 1958 with a degree in electrical engineering, he was drawn to audio. In addition to his busy recording career, he has spent 17 years as a professor at Goucher College and then at Johns Hopkins University.

Louis Mills: I think I wanted to be a recording engineer since I was 14, although at that point I wasn't exactly sure what one was. I cut my first commercial (for an automotive dealer, of course) when I was 16, on a wire recorder. Actually, I tied it as much as I cut it. I really didn't enjoy the summers I spent as an engineer, working in the pure scientific environment of the applied physics lab. I like being around people too much.

In the fifties I was recording music groups, concerts, choirs, and plays in stereo, then making my money by selling mono pressings for the groups since stereo records were a few years off. I had one of the first stereo Magnacord tape machines. People were pretty impressed by stereo back then. I still am. I started the recording studio in conjunction with WFDS in Baltimore in 1958. The station carried a classical format and operated from 6 till midnight, so that gave me the facility from 8 to 6. The station lasted less than a year and was acquired by WBAL. With both an AM and FM station, we were one of the first stations to broadcast AM/FM stereo. One channel would be broadcast on AM, the other on FM. Although WFDS did experimental FM stereo broadcasting during its short life, the FCC was trying to figure out which of several competing systems to go with. Our FM stereo system was better, but we lost. There was really no stereo program material at that time, so I had to go out and record it. I never had a mono tape recorder until a few years ago.

We added a record-cutting plant along the way and used a flour blender for mixing batches of vinyl. The money from those sessions and from the record-pressing business we operated made us enough to expand the studio.

By 1962 we had begun to syndicate commercials with Golnick Advertising. The idea was that if the commercial was written the right way, it could be used in a number of different markets. Well, the business just took off. So much so that it became a lot more profitable to do commercials than to do music. We had one of the first 1-inch Ampex eight-track machines. Because we weren't recording much music then, it went to Johns Hopkins Hospital. They used it for hearing research.

I later formed Images International, which specialized in syndicated TV spots. Now I'm back at Flite Three, pretty much where I wanted to be at 14.

With all of the different types of productions you do, what considerations make a production greater than the sum of its parts?

Louis Mills: One of the most important things is to make the technology as transparent as possible—both in the process and, even more importantly, before it gets to the actual process of a session. You've got to chew up and swallow the technology so that it's totally internalized, so that it is not an apparent factor in the session. I feel very strongly that your contribution should become as much an artistic one as it can possibly be. I think that's the difference between those who do this really well and people who don't. Anytime that the technology gets in the way—having to do things that take time or that take away from the focus of communicating the words off the page and into the listener's mind—it's a terrible waste.

Nelson Funk: It depends on the length of the piece. Being great for 30 seconds and being great for 30 minutes are two different ball games. As far as commercials are concerned, if it's not well conceived and written, it doesn't have a chance.

Suppose a client comes in and says,"OK, this is going to be an understated piece of work. We've got this guy who barely speaks above a whisper. We've got this real fine cello line that runs in the back, traffic sounds and people walking around looking at houses." You look at the copy, and it's almost a page long. Immediately, there's no way to establish this mood with this many words. The cello will sound out of place; the traffic will make the guy sound like he's rushing. If the pacing's off from the top or if the conception of the pacing is wrong, there's nothing we can do with music or sound effects that will heal this wound.

Let's sit down and see if we can get this copy down from 45 seconds to 20 seconds. Even though this is a 30-second spot, if we don't have 10 seconds to play with, none of this is going to work. If you've got a writer who knows all of this, he's probably working for an ad agency.

We may get involved with the casting. We may play the client as many as six different voices of guys who can really do this mood well. We'll choose a voice and record a voice track where the timing is good. You spend some time searching the music library for a few good cuts. Then we start building music and effects tracks.

Do you have any specific sequence for putting it all together?

Nelson Funk: Actually, the first thing I would do is find two or three pieces of music. That's to set the mood, the tone, and the pace of the whole thing. Then I'd work with them as they finalized the writing.

Louis Mills: Almost half the time I will have the piece of music chosen, as well as two others, before the session begins. The music choice is based on my knowledge of the client and on the nature of the spot—in that order. I know some clients who like electronic music; others don't like anything that wasn't played by a real instrument. If you know that, you can narrow your music library choices down by 50%.

I pick what I think they want, and then I pick what I think I would like to do if I went against the grain of the obvious. Incidentally, the third piece of music is always a "counter." I always have a piece of music that is counter to their concept of the copy. About one time in five it saves the day because the spot is not working the way it was conceived.

What do you look for in the copy to establish that counter piece of music?

Louis Mills: Say it's a car spot, and the copy's talking price, price, price, price, luxury, price, luxury, price, luxury, luxury. I'll pull a piece of luxury music as well as all the price music I think they'll want. You have to go through the final filter of knowing how far you can push the client.

Have you ever gone into the studio with an idea of what the production will sound like and had the whole idea change as you began to put it together?

Nelson Funk: Absolutely. Let's say we do the spot with no music, do all the sound effects, do the voice read, get the read that the producer has in his ear. Then we go in and pull four different pieces of music. Well, the whole structure of the thing can change.

If you have the music as an ear guide as you're doing the rest of it, the performer may give me something I can't direct out of him. If the producer asks for it, we can feed the music to the performer without recording it on the voice track, and that can significantly affect the read and give the talent a direction.

Louis Mills: I will almost insist on the talent hearing the music because in my mind there is no possible way to achieve proper focus without hearing it. I don't even split-track it. I mix on the fly. Once I've got the take I want— if I'm working on two-track machines—I'll pan the music to one side and the voice to the other and ask for one more take. That way I've got the voice track in the clear.

If I'm working on a piece with lots and lots of sound effects—not just background sounds, but door knocks, bells, and things like that—I'll mix the read and the sound effects hot so the talent can react to the sound effects. I'll also split the music from the voice/sound effects track so I can come back and do the moves on the music as part of the final mix. With the digital workstation, all that is easy to do.

Do all performers have an equal ability to react to the music?

Nelson Funk: No. That's why some performers make $200,000 a year, and some don't. If you've cast it properly and you're going into a session like this where it's constructed on paper, it's either going to become magical as a result of things that happen at that session, or it won't.

There's a big difference between on-air people at a radio station and people who do voice work for a living. Most radio personalities don't get better at doing voice work except by accident because no one critiques their performance the way they do a free-lance talent's, especially when it comes to production.

I've worked with some performers who first give the read the producer wants. After it's on tape, the performer will ask for another take. The performer knows he's got one more trick card he can play. When the performer saw the words, heard the music, and heard the rest of the spot, he's thinking, "I've got this thing I do that they don't know about in there." It also looks like he's helping to make the piece better. He'll do something totally different, and the producer will say, "Hey, let's play that back. It's not what I had in mind, but it's good."

What if the producer has a very limited window of acceptability?

Nelson Funk: If you know the producer is that kind of person, you don't offer to do that. If that guy works that way—and take 18 is the take he likes–with him this is good. Let's all get out of here before he changes his mind.

What are the benefits of collaboration?

Louis Mills: I have a special unlit talk-back button which allows me to let the performer hear what's going on in the control room, even though the client doesn't know it. While the producer is agonizing over trying to explain to me so that we can figure out what to tell the talent, I'll make the decision that this particular talent can handle whatever may be said. I feel that the chance of hurting the talent's ego is not nearly as harmful as having the talent not know what the producer has in his mind. I will mute the talent's mike so that he can't inadvertently say something to blow it, or I'll say something to the producer like, "They can't hear us, so we can talk freely." If it's all flowing out of the producer's mind and right out of the talent's mouth, none of this happens. Sixty percent of the time that doesn't happen, and somebody somewhere needs some help. Any help I can give them by any means, foul or fair, I will.

Nelson Funk: If you're in the right place with the right people, it should be a collaborative effort. There's a thing about the performer psyching up the producer as well as the producer psyching up the performer. If the performer has worked with the producer or can read the producer as being open to something else, it may work. As long as the producer doesn't feel threatened or offended or isn't made to feel that the performer is taking over.

One of the most frustrating things for a performer is to put him out in the studio where he can hear you and, because you operate the talk-back system, he can talk to you only when you want him to. You do a take and then say, "Do it again." He does another take, maybe three or four. Then when you ask him to do it again, you better start telling him why you want him to do it again.

A good performer will find a way to ask you that question. Is it the timing? You don't like the read? Am I hitting some wrong words? What is it? If you just keep doing the thing over, you're going to freak this performer out. You've got to be able to tell the performer what you want.

Louis Mills: I think it's absolutely incumbent on the engineer/producer to add to the creative process. If a director is searching for a way to communicate a thought to a talent I've worked with many times, I may know just the right thing to say to make the talent give the read the director is looking for but can't explain.

These days, the producer for an advertising agency may be an entry-level job. They don't always have the skills necessary to get the thing across. I have clients who are comfortable sending a junior producer to the session because they know I'll make up the difference. Not because of my "wonderfulness," but because I can figure out what they want and get it to the talent. Then the three of us can make it happen.

Nelson Funk: You can read producers who will accept input by the way they work with engineers. When the producer does something and turns to the engineer and says, "How do you like that? Do you think it's working?" he wants input.

It's a funny thing. A good engineer does this 8 hours a day. A good producer does two sessions a week. The engineer just might have something up his sleeve that would help, too. The smart producer calls on what he's paying you for. He's paying the performer for not just his voice, but his mind. He's paying the engineer in the studio to help him, too.

That doesn't mean he's going to fax over the copy and let us produce it for him, although we do that, too, sometimes. If there are formula spots week after week, instead of coming to the studio, the client may direct the session by phone.

Louis Mills: Fax machines have changed the world. Fifty percent of the copy that's done here is faxed in before the session. I take a quick look at it and make most of the decisions I'm going to make, short of talking to the client during the session, in that one glance. I make a lot of mental notes at that point. I can't tell you how many people come to a good recording engineer because that person makes them feel more secure. Even if making them feel more secure has nothing to do with the copy, the concept, the music, or anything else.

Then what you're saying is that, even for the engineer/producer, there is a performance aspect?

Louis Mills: Absolutely! And it gets back to what I said about swallowing the technology. Get this equipment out of here. Who cares? We're talking about *art!*

I'll sometimes jump in and suggest a different read to the performer, which a lot of people think is audacious. Then, too, many agency producers are very insecure. If you're good enough, you can help them because they know I've been doing this a long time and that I might just have a good idea. I was once told that I was swimming in a sea of narcissistic self-gratification. I'm not sure that's all bad. I have the opportunity of working with someone who comes in with nothing. Two hours later they leave with a shining little masterpiece. That's my job.

There's always somebody responsible for the temperature of the session. I feel it's my job to do that unless somebody has clearly taken the role.

What can you do if you become aware that the session is going badly because of some emerging bad attitude?

Louis Mills *(Throwing a magic marker at the interviewer):* That! I do something or say something that's absolutely stupid or outrageous, and then laugh. But I'm dead serious when I say something, and they're not quite sure. Most of the time, not being sure, they'll lighten up. A little bit of shock or surprise will help get them back on track. If they come in loaded down with problems, I often address that with them in an obtuse sort of way.

What are the major mistakes that occur during a session that keep the best work from being done?

Nelson Funk: We have some young, inexperienced producers who start the session by asking how long they have the performer. An hour and a half if you're going to do one spot [according to American Federation of Television and Radio Artists (AFTRA) commercial radio codes]. They say, "Great. We'll just keep doing this for an hour and a half." At that point, I tell them it'll probably get worse if they do it that way because there's no reason for it to take an hour and a half to get the voice track. It might take that amount of time to edit the voice, music, and sound effects and do a mix. You shouldn't want this person to read this 30-second piece of copy for an hour and a half because it ends up sounding really bad. In addition, you end up listening to 40 takes only to go back and use take 3, which was the good one.

Louis Mills: Not recording the first read! Man has but a few magic moments, and that first read just might be *it*.

How about starting the session with copy that's too long?

Nelson Funk: That's another thing that can drive you crazy. A guy comes in with copy that's looks 40% too long. Then he says, "Look, I don't want you to pitch this. I want it laid back." You're looking at 45 seconds of copy, and you want the talent to talk slow?

The first read proves it out. The producer cuts out three words and says, "OK, let's time it again." Well, now it's 35% too long. "OK, let's drop the *the* and the *but.*" Finally the engineer may say, "We need to cut three lines here, cutting *but* and *the* isn't going to make it."

Meanwhile, the performer has now read 15 takes knowing he hasn't a prayer of getting it into a usable time frame. So, by the time we get through this whole exercise, it's really stale; he's read it too many times. And we're just getting into interpreting and directing because we've spent 45 minutes editing the copy. That can mess up a session quicker than anything I've seen.

The producer will say, "I don't understand this. I read it in the office and it only took me 30 seconds." Of course, he didn't read it out loud; he read it with his eyes.

What's the best-case scenario?

Nelson Funk: The progression of events has a lot to do with whether it comes together. Say you've got a concept for a spot. You don't have enough money for an original score, so you do a music search to come up with several pieces which will work.

Then you go back and write the spot. Now you're hearing the music as you write the spot, and you're timing the copy to the music. So you have the mood established, you have the music, the copy has been written to time, and now it's time to concentrate on the next element—the voice track. You get a really good voice reading: the right timing, the right pacing, the right feel. Now, if there are effects involved, it's a matter of sliding around the voice track and building the elements so the right effects hit in the right holes, the mood is right, and it's seamless. Then you do a mix.

The mix, of course, can be very important. Again, is the mood working? Do the effects work with the music? Was the read right now that you've heard all the elements together? The chances are it's all going to work because it started with a good idea that was firmly in place, and the producer is now hearing what he wanted to hear. It's come down painlessly for everyone. He hasn't burned a lot of studio time, he hasn't burned out a performer, and he didn't have to make 145 takes to get what he wanted.

Louis Mills: Writing to a prechosen music bed is ideal, but from a practical standpoint, it seldom happens. Often, there's not enough time to do a music search in the session, but when it does happen, it can be great!

I think the handling of the music is the greatest art there is—how you do a hot mix or a cold mix. Where you get that music up and how conscious you are of everything that's going on at the same time, while having the music being as supportive as possible, is the key. I'm constantly moving the music levels. I can do the music changes blindfolded and in my sleep after I've heard each piece twice. I only know it for 30 minutes, and I'll have to re-learn it if I use it 2 days later. After a while, my third ear automatically tells me to bring up the gain of the music. I don't change voice levels unless it's a pretty spot and I have to bring up things that are missing or if the talent really doesn't maintain the proper level.

How and when should a producer determine how large or intimate a production should sound?

Nelson Funk: Even before you hire the performer. If the producer says, "OK, I've got this spot that takes place in the showroom. I want to imply that thousands of people are showing up to buy these cars. I want to start off in the showroom, then I want to hear all these brakes squeaking with the customers pulling up in their old, beat-up cars, and I want to hear the crowd getting bigger in the showroom, the announcer starts to get frantic because we can't get any more people in here. That's the mood I want."

You start out knowing you can use a lot of words because the performer will start slow and talk faster and faster. He'll get away with it because it's justified by the excitement and the energy and the things that are taking place. You have made that picture in the spot.

We're going to need tons of sound effects. It can't sound like one brake squeal looped and punched in a bunch of times. These are all different cars coming up. Guys in trucks, people in cars. They've got to sound like old beat-up, bad cars and trucks. Then we've got this huge crowd thing, and of course, as in most car spots, we may have a jingle tag sing at the end. As the music starts to build, more people get into the showroom. Meanwhile, the performer's got 45 seconds of copy he's going to do in 30 seconds.

The concept there was in place before we started. We all had the picture. The dead giveaway with a radio script for me, and we get this from young producers all the time, is a three-paragraph description of the scene at the top of the script. But nowhere in the body of the copy is any of this justified or explained. The copy never says we're in a car dealership. A savvy writer includes those kinds of things in the copy. Like, "We're here at XXXX

Ford this afternoon, and you won't believe what's going on out here, folks. We're in the showroom with all these cars and trucks, and people are driving up by the thousands to trade in their old cars and trucks."

Now if you didn't say something like that and you just put the brake squeals and other sound effects in, the listener would have a hard time telling what was going on. You haven't got a radio spot. Young writers spend so much time with TV that it doesn't occur to them that none of this will work.

Because in their minds they are seeing it as a TV spot?

Nelson Funk: Exactly. In fact, we've done spots where they have literally lifted the TV audio track for use as a radio spot. When it works, it works. When it doesn't work, they'll say, "Well, they're going to see it on TV and make the connection." I tell them I certainly hope they see it on television before they hear it on the radio because on the radio it doesn't make any sense.

How do you establish the size and nature of the space in a production?

Nelson Funk: You start by playing them a lot of sound effects and asking if it sounds like the space they're looking for. One of the strangest things about all that, and especially in long-form projects like films and videos, is that people go right to a location and record the actual ambience of, say, a car showroom. They bring it in, and during the mix you say, "What is that?" They'll say, "Oh, that's the real thing. That's the sound of the showroom." Fine: a big empty space with an air conditioner humming loudly in the background. It doesn't sound like a showroom. Then I play them a stock effect, which they like because their mind tells them that's what a showroom should sound like. If you're setting space in radio, you have to set what we've told them it is.

The great standard condo spot is another example. We're always doing this spot. It's a wonderful place to live. We've got gardens and walkways, and you hear birds. We have this lake where you fish, and you hear the fishing rod. You hear crickets at night showing how quiet and peaceful it is. You go inside, kick in the reverb, and talk about the cathedral ceilings. This is a standard spot, but we've told them about the scene before we let them hear it.

The reverse of that is to let them hear it first, and then explain where they are. It's a matter of design.

Louis Mills: It's got to come from the actor, too. If the actor's not doing it and hasn't been asked to do it, there's not much the engineer can do to create a believable sense of space. The experienced actor approaches the performances after answering a lot of questions. Questions like: Who are you talking to? How close are you? Who's the other person or persons you're talking to? Once the talent focuses on who they are, where they are, and who they're talking to, the rest becomes so much easier.

How does this affect how you set up for a session?

Louis Mills: You've never seen a studio I've worked in that doesn't have gobos, or movable acoustic flats. I use them to help create the size of the space. I'll also move the microphone in the studio a lot to get the right sound.

How far can you stretch the listener's imagination?

Nelson Funk (Laughing)**:** That's what we have producers for. That's a judgment call of the producer. The most important thing is how far you can stretch the imagination of the person signing the check for all this. Everything that gets produced gets played for someone who may be the least qualified to judge it, but that person has an image of what that place should be. If you don't please that person, it'll never get out of the studio.

It's a balance. How good is the relationship between the producer and the client? Does the client trust him? Are they buddies? Do they play golf together? Is it the kind of relationship where the client says to the producer, "OK, if you say so, Charlie. I don't like it, but you know advertising."

Maybe it's "I don't know. I'll take it home and play it for my wife. If she doesn't like it, it's out." That's usually fatal. I've had a client bring his kids into a jingle session. If they didn't tap their feet, it was out.

Other than music, how fully do you use stereo?

Nelson Funk: We do everything in stereo. We certainly don't ping-pong it—with one exception. I have done one commercial where the whole idea is the woman is talking to two different people. She's talking to people who want to buy a home, and people who don't. We start the spot with her saying, "All you people who want a buy a home, listen here on the left (voice pans to left). You people who don't want to buy a home, don't pay any attention to this. We're going to put some music over here (voice pans right) on the other side for you to listen to." She would pop over on the music side of the commercial from time to time to tell the nonbuyers that they were really missing something, but to stay there if they really didn't want to buy a house.

The commercial was tremendously successful, and they repeated the format for additional spots. One of the most interesting aspects, however, was that the initial buy relied heavily on AM. They got no complaints. Everybody understood it without it going from left to right, because it was done right. Because the copy was written so that she told the listener what side she was on, and when she moved, it worked. It was a prime example of theatre of the mind

Other than rare occasions like that, we don't pan things too much. In fact, we'll reduce the stereo separation on a lot of the music libraries. Depending on the composers, some of the elements are panned too radically to the left and right.

Louis Mills: I believe there are more people listening in stereo than we think. If you don't ask every client, as a PR move (to the days when we all will be doing Dolby Surround Sound), then their awareness will not be built as rapidly as it should be. I never pan the voices out more than 20 or 30 degrees, but as a rule I'm not bothered by music with lots of separation because the melody is seldom panned out that far. Most of the music is done so that even if you only hear one of the stereo channels of music, you're not in bad shape.

Any words of wisdom about the use of comedy?

Louis Mills: I think comedy is in a state of sad decline at the moment. Good comedy, such as in Dick Orkin and his Radio Ranch, is still the most effective thing. People talk about bizarre and comedic pieces; they don't talk about how neat a piece of music was. I'm a big comedy fan. It gets my attention, and it works if it's good.

Nelson Funk (Laughing): Get a writer who knows how to do it. A lot of things have changed in comedy. We used to do an Italian accent, a German accent, and you made the cliché of the guy who can barely speak English, and it was funny. It is no longer funny because there's no one left you can get away with offending. So you can't use classic cliché humor. If I make the guy with the accent the hero and you're an American and you're the dummy, it can work.

Is the craft necessary to produce a spot like that something you're born with, or can you learn it?

Nelson Funk: You learn all that by working with the person who produced the spot. You try to get work as a production assistant with people who are doing this kind of work. If you're really smart, you begin to understand what it is they do.

How much is inherent in a good producer, and how much can you refine or teach someone to improve their ability?

Nelson Funk: It's really hard to say. In long-form television, you can take three producers out and let them shoot the same story. Let's say it's an interview with a nationally known boxer. The best producer may take the guy, put him in trunks, put him in the ring, have the interviewer sitting in the ring with him, have the light pouring over the boxer's shoulders. You're looking at a beautiful portrait of a guy who makes his money with his body. This is where he works, but he's having an intimate conversation in this environment. Another guy might put him in a room somewhere or in a studio with two chairs and just shoot it like that.

If you turned on the TV in the middle of the thing, without knowing who either of these people were, and saw the interview in the ring, you'd say, "This guy is a boxer." If you saw the other setup, you'd say, "This guy's interviewing a boxer." Visual orientation is very important.

Is there any truth to the idea that some people are born with a natural aural acuity that gives them an advantage in the studio?

Nelson Funk: It can be an advantage and a disadvantage. I work with people who walk into a room with a video monitor on, and the sound from the horizontal oscillator drives them crazy. I don't know if that makes them better. So much of this is a matter of taste. That's why some producers have favorite engineers.

Louis Mills: I think it's only a slight advantage. Timing and acuity, although heavily related, are two different things. You can develop each of them, but I'd hate to try to develop them both at the same time. Timing is something you have to learn to feel; it's visceral. It's the degree to which you move on your chair as a listener, or maybe your head imperceptibly moves forward or backward to catch what's going to be said next. That's all timing. You can take the worst copy and make it work with timing.

If you asked me what to do to improve your timing, I'd have to say listen. I do an awful lot of listening, not only to performers in the studio, but everywhere—in the line at the butcher shop. I enjoy listening to people and music all the time.

Does all of this listening create a reference for the reality you might aim for in a particular production?

Louis Mills: That or an accentuated reality or a deliberately reduced reality. Lorenzo Music, who does the voice of Garfield, has an excellent sense of reverse-stretched reality. It's so offbeat and unexpected in the way that it flows that it forces you to pay attention to it. In addition to everything else, he has an interesting voice and is saying interesting stuff.

How much depends on what kind of system the finished production will be played on?

Nelson Funk: That's one of the big mixing problems. We're all learning with the client's money. You do a job; you hear it on the air. If you hear it on three stations and you think it's a little too bright on all of them, it's a little too bright. If you remember what it sounded like, next time you'll back off a little bit. You're tuning your craft as a result of where it goes. Radio and television is relatively easy to get a fix on. You can hear the results.

Louis Mills: I ultimately bought a set of speakers for my home that replicated, as closely as possible, the monitors in my studio because I wanted to hear what things sounded like on radio and TV over roughly the same speakers in two different places. I normally use three monitor sources in the studio: the big point-source ones, a set of near-fields, and the small one in the meter bridge of the tape machine. I'll even take a chance and make a cassette dub so the client can listen to it in the car.

I listen to radio and TV all the time. My car has the controls built into the steering wheel. I station hop for commercials constantly. I do it at home the better part of an hour a day, when I'm doing other things around the house. The experience of listening, the memory that I've developed over years of listening and making decisions, also plays a major part in my judgment.

Nelson Funk: Suppose a producer wants to do a piece that will play in a big museum somewhere. They may have gotten an original score done with a huge arrangement that really ties in with the visuals.

We always ask all the questions we can think of that will affect the audio: How are you going to play it back? How big is the room? How close are the people to whatever you're playing it back on? Will there be 40 or 400 people in the room? Then we do a mix. We listen on good speakers and bad speakers. It's all a set of compromises. The producer is happy and leaves with the mix.

Invariably, you get a call from an unhappy producer in a day or two. So you go to the museum, and lo and behold you find a home hi-fi system hooked up to a quasi center-channel amplifier. They have put in some sort of quasi surround system, and all of it is coming out of really bad speakers.

Incidentally, this is a marble room with a 50-foot ceiling. You sit in the audience area and the voices from the people on the screen are coming over your right shoulder because there's a speaker on a pole over there somewhere. You suggest that they cut the side speakers. They say they're trying to get "surround sound." They're getting it, but at the price of intelligibility.

With the technology as advanced as it is today, it's not a crazy idea to go into that room and mix it there. With a number of pre-mixes laid up in the right manner and with a portable mixer, you can do the real final mix right in the room over the system it's going to be played on. Of course, if they move the exhibition down the hall or send the tape out to be played somewhere else, it'll probably sound pretty strange.

What are some of the other factors that a good producer takes into consideration?

Nelson Funk: Something as simple as the level at which a mix is played makes a tremendous difference in the relative levels of the elements of the mix. Five decibels of overall level difference in playback level makes a big difference in the way a mix sounds because the ear is not linear.

In theory, voicing a room is a good idea. However, you won't know you've got a problem until the audio at a certain frequency hits some level and becomes resonant. If you know what level you're going to play the mix, you can voice the room at that level and make any adjustments. If you exceed that level, you may run into problems you never knew existed. Too high or too low—both can be tough problems.

I have played back things in rooms where it turns out the narrator is talking too fast for the reverb time in the room. If we had slowed the narrator down, the intelligibility would come way up. At the faster speed, the reverb in the room is causing his voice to interfere with itself. A slower pace would have been better. So, are you going to tell your next client you want to do a real-time analysis on the room where this thing will play so you can pace the narrator? No way!

Louis Mills: It's also a good idea, when creating ambient sound, to make it thick in texture but low in level. In other words, use plenty of it to make all the elements that might be there be there. If we want to be outdoors, we want some traffic, some birds, some crickets, and some wind, whatever. Make it thick, but don't make it terribly loud. The mind takes care of that beautifully. The thicker you are with elements, the more intrigued the ear becomes.

Then, too, listening with an innocent ear after you've heard something 67 times is a tremendous art and hard to develop. At some point, you may have to stop and say, "Wait a minute. I'm not listening right," correct yourself, and go back and listen again.

How has your use of compressors and limiters been affected by the loudness wars in radio and TV?

Nelson Funk: This is one of the things that separates one studio from another. If my client remarks about how the audio on one of our spots really jumps out, I've got a happy client. This is a big commodity for me to sell. If his spot ends up next to a national McDonald's spot and his local car commercial sounds the same—the level, the crispness, the brightness, the ratio of music to voice—if it doesn't fall apart upon direct comparison, he's happy. That's what we try to do.

Louis Mills: I think I have managed to make my commercials sound louder than other people's commercials. That's my ultimate technical goal.

To be able to hear a spot in the kitchen is the thing I worry about most. Of course, if there's any socially redeeming value and something like art present, that's real nice too.

Do you have any recommendations for using limiters and compressors?

Nelson Funk: I don't want to talk too much about this. Our selection of compressors and limiters is our own business. Every studio has their own way of processing a mix. The only thing I will tell you is that we try to send our work out of here so that the station's processing gear won't kick in. Do we cut off everything below 50 Hz for TV and FM? No, we don't. We also don't send out stuff that's got 0-dB level at 20 Hz. When we hear something like that, we know it's just going to decrease the overall apparent level of the spot, and no one's going to hear it anyhow.

I once heard some stuff I thought was incredible; we air-checked it and brought it into the studio. When we played it in our studios, it sounded awful. Our clients would never let us do this. This client let the producer do his thing. He must have had every box in the room lit up. I saw the spot in the middle of a bunch of other national spots, and the thing really jumped off at you. The voice was so sibilant it could cut glass, but it didn't sound bad on the air. All the processing that was added on by the TV station was also a factor, but I've heard this spot on several stations, and it was mixed that way.

Louis Mills: We do use a few black boxes in the chain, which we just don't discuss, but I can say that my mixes for radio and television are clearly hotter in the upper midrange—around 8 kHz. And mixes for radio and television are different. I've found that in order to stay out of the way of the station's own processing, I can roll off at 50 Hz and a little at 13 kHz for both AM and FM. Then I do all my brightening and upper bass stuff so that all the other things can pop through. That way I'm somewhat protected from what the stations do, and I get more energy in the bandwidth I use. I never boost low bass—midbass yes, but not low bass. I cut lows on TV even a bit more because until more people are listening over good monitors, they'll never hear some of the stuff we could give them. But if I do it right, A/B comparisons between my mix and a conventional mix, on good speakers, sound pretty close.

I also limit and compress before and after EQ. There are probably not ten studios in America like this, but I run a Urei 1176 as part of the "normalled" output of the console. I compress 4:1 a fair amount, and I limit 20:1 just enough to knock the tops off. I run my 4:1 compression slope pretty heavy, rather than an 8:1 or 16:1 or more. I want my four working a lot. I can slip more stuff through the station sound chain that way. Depending on the job, I may use from 4 to 10 dB of gain reduction at that 4:1 slope. I'm packing the voice, sound effects, and music tracks together with that much compression after I've packed them going into the mix. Furthermore, if I'm doing it in stereo, I'm deliberately doing crazy things with the left and right ambience—particularly in the holes of the piece—so that it becomes noticeable if you're listening for it. By filling up spaces like that, you're adding listening interest, even if you're doing it in mono. It makes the sound become more arresting to the ear, which is what counts.

I do believe in reverb. I believe in fattening the track much the same way the Haas effect fattens a track. Not the same way, but in a different way. Just enough reverb so that you don't really hear it, but it's there, and it affects you.

I sort of know in my mind how to circumvent the EQ, limiting, and compression a station may use, but I have a real problem with aural exciters. There's no way I can compensate for their effect when I'm recording. All too often, exciters produce a lot of sibilance. Bite's fine, if it's that type of commercial, but not splattering sibilance, which may start when the spot is dubbed to cart and then processed through the main audio chain. There's nothing I can do about that.

To what ends do you go to ensure the audio quality of your dubs?

Nelson Funk: In TV, many of our clients bring their 1-inch videotapes to us—not the masters, but the dubs. We do a layback from our mix to all of the dubs. That way, the audio is only one generation down from the audio master. We've heard a tremendous improvement in the quality of our audio on the air as a result of doing it this way.

Louis Mills: If I'm mixing a stereo commercial, I'm not going to make any assumptions about the equipment it's going to be played back on. First of all, I check very carefully to see if the station is really going to play it back in stereo. That may change from week to week.

If they're doing a stereo television commercial, I do my damnedest to make it sound superb in every sense. I don't screw around with it too much, trying to get around other devices. I've got to make an assumption that somebody's going to do something right somewhere. If it's not a stereo commercial, I try to put in the filters in my head that create a mix that will average the best sound for all of the stations.

How about audio processing for long-form projects like TV shows?

Nelson Funk: You get in big fights because in half-hour shows clients want to create some very subtle moods. Our monitors are so good in the mix rooms, you can do this. Producers want to do things down at −25 dB. I say, "Look, there is no −25 dB. We have it here, you're hearing it, and you're liking it, but on the air there's no −25 dB. They don't see that number; they take care of it. I'll tell you where they'll pull it. If you don't let me put it down right, it's not just the traffic that will get pumped; everything will get pumped. And when it gets pumped and you hear a word of dialogue, the dialogue will get squashed." It's a bitter fight because it destroys the event. Everybody's got this good feeling about the mix, and I come along and tell them it won't make it to the TV speaker.

We never put music, sound effects, dialogue, or studio voice tracks through the same limiter set the same way. We don't limit the mix; we limit the elements as they're being mixed. What's a perfectly good ratio, attack and release time for a dialogue track isn't going to work for a piece of music. It takes a lot of limiters and compressors. What we're doing is getting our total program content to look like something their transmitter processors don't want to smack up or pull down.

What about location audio?

Nelson Funk: That's the hardest. You get well-recorded tracks, but in tough situations. The audio often requires a lot of EQ, processing, and noise reduction. When you've spent a day mixing one of these programs, you may feel very unrewarded. When it's over, you say, "I really didn't make this sound great, but I think I made it as good as I can get it without having everybody sound like they're talking on the telephone." You can increase the intelligibility and destroy the character of the voice. The producer has to make the judgment.

So about a week later, you see and hear the piece on the air. I have actually called clients and said, "Did you guys mix that over? It sounded so good on the air," only to find that it was our mix.

What do you do about the inevitable generation loss?

Nelson Funk: Take a piece that's shot on film. They used a Nagra to record the audio; they transferred the audio to 16mm or 35mm film. We mix it to another 35mm. We laid it on 1-inch videotape—that's generation 4— and it's laughingly called the *master*. Generation 4 goes up to the network and is integrated into the program reel, that's generation 5. That tape and sometimes a backup roll on the air. That means you're listening to generation 5 or 6 before it hits the network line audio processing. Then it comes down to the local station, which processes it again before it is broadcast. That's what you're hearing on the air.

We don't have a problem with tape noise. Everything is Telcom or Dolby all the way down the line. Five generations of Telcom with 118-dB signal-to-noise ratio is fine. It's the number of limiters and compressors and equalizers that have squashed the audio. At some point, it's all out of your hands anyhow.

Suggested Reading

Alten, Stanley R. *Audio in Media*. 3d ed. Belmont, CA: Wadsworth, 1990.

Bartlett, Bruce. *Introduction to Professional Recording Techniques*. Indianapolis: SAMS and Co., 1988.

Borwick, John, ed. *Loudspeaker and Headphone Handbook*. Boston: Focal Press, 1988.

Gross, Lynne S., and David E. Reese. *Radio Production Worktext: Studio and Equipment*. Boston: Focal Press, 1990.

Huber, David Miles. *Audio Production Techniques for Video*. Boston: Focal Press, 1992.

Huber, David Miles. *Microphone Manual: Design and Application*. Boston: Focal Press, 1992.

Keith, Michael C. *Broadcast Voice Performance*. Boston: Focal Press, 1989.

Maestas, Bobby. *Recording Sessions*. Newbury Park, CA: Alexander, 1989.

Martin, George. *All You Need Is Ears*. New York: St. Martin's Press, 1979.

Milano, Dominic, ed. *Multi-Track Recording*. Milwaukee: Hal Leonard Books, 1988.

Nisbett, Alec. *The Use of Microphones*. 3d ed. Boston: Focal Press, 1989.

Oringel, Robert S. *Audio Control Handbook*. 6th ed. Boston: Focal Press, 1989.

Pohlman, Ken C. *Principles of Digital Audio*. Indianapolis: SAMS and Co., 1988.

Thom, Randy. *Audio-Craft: An Introduction to the Tools and Techniques of Audio Production*. 2d ed. Washington, DC: National Federation of Community Broadcasters, 1989.

Watkinson, John. *The Art of Digital Audio*. Boston: Focal Press, 1988.

Index

A

Acoustic noise, 26
  causes, 26
  instrument cable, 26
  microphone cable, 26
Acoustical joining, 18
Acoustics
  control room, 15–17
  live end-dead end (LEDE) theory,
    16–17
  studio, 15–17
  surfaces, 16
  volume, 17
Additive equalization, 67–68
Adjustable equalization
  continuously variable, 63–64
  step-switched, 63–64
Adjustment
  expander, 81
  gate, 81
Advertising agency producer, 125, 126
Alternate main output, 59
Ambience, 88–89, 129–131
Analog cart machine, 113
Analog equalization, 63–68
Aphex Aural Exciter, 92
Attack time, 75–76
  compressor, 75
  limiter, 75

  transient, 75–76
Audible spectrum, 2
Audio logic processor, 92–93
Audio processing, 62–94
Audio-processing circuit, 61
Audio signal flow, 45–60
  diagram, 45–46
Audio switcher, 56–57
  memory, 56–57
Auxiliary send, console, 58–59

B

Balanced audio, 40–43
Bandwidth control, 71
Bass frequency
  equalization, 6, 7
  monitor, 19
  proximity effect, 69
  voice track, 69
Bass trap, 16
BBE Sonic Maximizer, 92
Bedini Audio Spatial Environment
  processor, 93–94
Bit rate, digital audio, 99–100

C

Cable, polarity, 54–55
Cannon connector, 40, 43
Capacitance, 33–34